Introduction to ENVIRONMENTAL TOXICOLOGY

Introduction to ENVIRONMENTAL TOXICOLOGY

Volume–1

Patricia Shaw
Diana Hutchinson

Kruger Brentt
Publishers

2024

Kruger Brentt Publishers UK. LTD.
Company Number 9728962

Regd. Office: 68 St Margarets Road, Edgware, Middlesex HA8 9UU

ISBN: 978-1-78715-128-4 (Vol. 1)

For information on all our publications visit our website at http://krugerbrentt.com/

PREFACE

Environmental toxicology is the study of the effects of chemicals and other substances on living organisms and their environment. It is a multidisciplinary field that combines elements of toxicology, environmental science, and ecology. The impact of toxic substances on the environment has become a growing concern in recent years, and the field of environmental toxicology plays an important role in assessing and mitigating these effects. This essay will explore the field of environmental toxicology, its importance, and some of the challenges it faces. Environmental toxicology plays a critical role in protecting human health and the environment. Toxic substances can contaminate air, water, and soil, and can have serious consequences for both human and animal health. By studying the effects of these substances, environmental toxicologists can identify potential hazards and develop strategies to mitigate their effects. Environmental toxicology is also important for regulatory agencies that set standards for exposure to toxic substances. These standards are designed to protect human health and the environment, and are often based on the results of environmental toxicology studies.

One of the major challenges of environmental toxicology is the complexity of the environment. The effects of toxic substances on living organisms can be influenced by a wide range of factors, including temperature, humidity, and other environmental conditions. This complexity makes it difficult to predict the effects of toxic substances in different environments. Another challenge is the lack of information about the effects of many chemicals on the environment. There are thousands of chemicals in use today, and only a fraction of these have been thoroughly tested for their effects on the environment. This lack of information makes it difficult to assess the risks associated with many chemicals. The future of environmental toxicology will be shaped by advances in technology and a growing awareness of the importance of protecting the environment. New tools and techniques, such as computer modeling and high-throughput screening, will allow researchers to study the effects of toxic substances more efficiently and effectively. These advances will also allow researchers to better predict the effects of toxic substances in different environments. Environmental toxicology is a critical field that plays an important role in protecting human health and the environment. While the field faces many challenges, advances in technology and growing awareness of the importance of environmental protection offer opportunities for progress. By continuing to study the effects of toxic substances on the environment, environmental toxicologists can help ensure a safer and healthier world for future generations.

The present book provides a good general introduction to all the major areas of environmental toxicology,

Fundamental aspects of environmental health, occupational health, detection of pollutants, and risk assessment are discussed. This book provides a thorough, systematic introduction to environmental toxicology and addresses many of the effects of pollutants on humans, animals, and the environment. The book will be of interest to biologists, environmental engineers, chemists, environmental scientists, microbiologists, medical doctors and all academics, professionals, policy makers and practitioners involved in the environmental toxicology and hazardous waste management.

We are grateful to all those persons as well as various books, manuals, periodicals, magazines, journals etc. that helped in the preparation of this book. In spite of the best efforts, it is possible that some errors may have occurred into the compilation and editing of the book. Further queries, constructive suggestions and criticisms for the improvement of the book are always welcomed and shall be thankfully acknowledged.

Patricia Shaw

Diana Hutchinson

CONTENTS

Chapter 1 ENVIRONMENTAL TOXICOLOGY

1.1. ENVIRONMENTAL TOXICOLOGY

Environmental toxicology is the science that studies the fate and effects of potentially hazardous chemicals in the environment. It is a multidisciplinary field assimilating and building upon knowledge, concepts and techniques from other disciplines, such as toxicology, analytical chemistry, biochemistry, genetics, ecology and pathology. Environmental toxicology emerged in response to the growing awareness in the second part of the 20th century that chemicals emitted to the environment can trigger hazardous effects in organisms living in this environment, including humans. Section 1.3 gives a brief summary of the history of environmental toxicology.

One way to depict the field of environmental toxicology is by a triangle consisting of chemicals, the environment and organisms (Figure 1). The triangle illustrates that the fate and potential hazardous effects of chemicals emitted to the environment are determined by the interactions between these chemicals, the environment and organisms. The fate of substances in the environment is the topic of environmental chemistry, the effects of substances on living organisms is studied by toxicology, and the implications of these effects on higher levels of biological organization are analyzed by the field of ecology.

Another term widely used to refer to this field of study is ecotoxicology. The main distinction is the inclusion of human health as an endpoint in environmental toxicology, whereas ecotoxicology is restricted to ecological endpoints. Since the current book includes human health as an assessment endpoint for environmental contaminants, the term environmental toxicology is preferred over ecotoxicology.

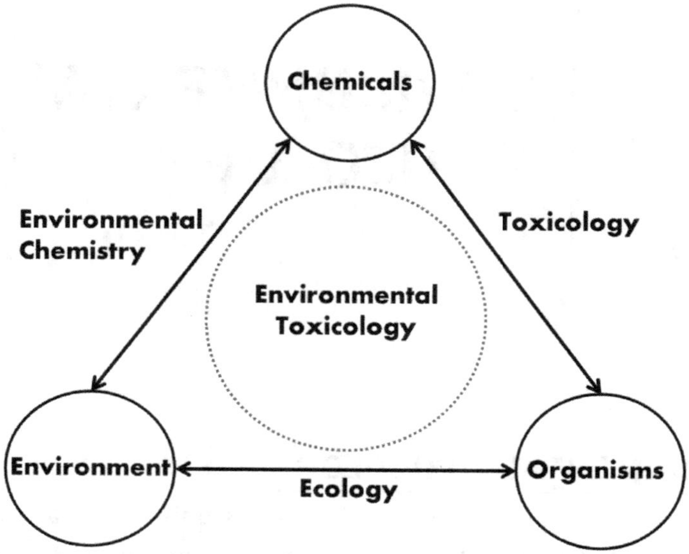

Figure 1: Environmental toxicology studies the interactions between chemicals, organisms and the environment making use of environmental chemistry, toxicology and ecology. Source: Ad Ragas.

Environmental chemists study the fate of chemicals in the environment, e.g. their distribution over different environmental compartments and how this distribution is influenced by the physicochemical properties of a chemical and the characteristics of the environment. They aim to understand the pathways and processes involved in the environmental fate of a chemical after it has been emitted to the environment, including processes such as advection, deposition and (bio)degradation. Within the context of environmental toxicology, the ultimate aim is to produce a reliable assessment of the exposure of organisms, an aim which is often complicated by the enormous heterogeneity of the environment.

Environmental chemists use a variety of tools to analyze and assess the fate of chemicals in the environment. Two fundamental tools are analytical measurements and mathematical modelling. Measurements are essential to acquire new knowledge and insight into the behavior of chemicals in the environment., e.g. measurements on emissions, environmental concentrations and specific processes such as biodegradation. These measurements are analyzed to discover patterns, e.g. between substance properties and environmental characteristics. Once revealed, such patterns can be integrated into a comprehensive mathematical model to predict the fate of and exposure to substances in the environment. If sufficiently validated, these models can subsequently be used by risk assessors to assess the exposure of organisms to chemicals, reducing the need for expensive measurements.

Chapter 2 focuses on the types of chemicals occurring in the environment, their sources and the concentrations found at contaminated sites. In Chapter 3, focus will be on the fate and transport of these chemicals, including aspects of bioavailability and bioaccumulation in organisms.

Toxicologists study the effects of chemicals on organisms, often at the individual level. Fundamental toxicologists aim to understand the mechanisms involved in the toxicity of a compound, whereas more applied toxicologists are primarily interested in the relationship between exposure and effect, often with the aim of identifying an exposure level that can be considered safe. Within this context, the dose concept as introduced by Parcelsus at the start of the 16th century is essential (see Section 1.3), i.e. the likelihood of adverse effects depends on the dose organisms are being exposed to.

The processes taking place after exposure of an organism to a toxicant are often divided into toxicokinetic and toxicodynamic processes. Toxicokinetic processes are those that describe the fate of the toxicant in the organism, including processes such as absorption, distribution, metabolism and excretion (ADME). These toxicokinetic or ADME processes are sometimes collectively referred to as "What the body does to the substance" and determine the exposure level at the site of toxic action, or internal dose. Toxicodynamic processes are those that describe the evolution of an adverse effect from the moment that the toxicant, or one of its metabolites, interacts with a molecular receptor in the body. This interaction is often referred to as the primary lesion or molecular initiating event (MIE). Toxicodynamic processes are sometimes collectively referred to as "What the substance does to the body" and the chain of events leading to an adverse outcome as the adverse outcome pathway (AOP).

The toxicity of a compound thus depends on toxicokinetic as well as toxicodynamic processes. Traditionally, this toxicity is being determined by exposing whole organisms in the laboratory to the substance of interest, and subsequently monitoring the health status of these organisms. However, as a result of the growing societal pressure to reduce animal testing, as well as the increased mechanistic understanding and improved molecular techniques, this so-called "black box approach" is more and more being replaced by a combination of in vitro toxicity testing and "in silico" predictive approaches. Physiologically-based toxicokinetic (PBTK) models are increasingly used to model the fate of chemicals in the body, resulting in a prediction of the internal exposure. In vitro tests and advanced molecular techniques at the gene (genomics) or protein (proteomics) level may subsequently be used to determine whether these internal exposure levels will trigger adverse effects, although many challenges remain in the prediction of adverse effects based on in vitro test and omics information. Chapter 4 focuses on dose-response relationships, modes of action, species differences in sensitivity and resistance against toxicants.

Ecologists study the interactions between organisms and their environment. Ecology is an important pillar of environmental toxicology, because ecological knowledge is needed to translate effects at the individual level to the ecosystem level; an important endpoint of ecological risk assessments. Such a translation requires specific knowledge, e.g. on life cycles of organisms, natural factors regulating their populations, genetic variability within populations, spatial distribution patterns, and the role organisms play in processes like nutrient cycling and decomposition. Effects considered relevant at the individual level, such as a tumor risk, may turn out to be irrelevant at the population or ecosystem level. Similarly, subtle effects at the individual level may turn out to be highly relevant at the ecosystem level, e.g. behavioral changes after environmental exposure to antidepressants which may affect the population dynamics of fish species. In recent years, there is an increasing interest for the role of the landscape configuration, distribution patterns and their dynamics in environmental toxicology. The spatial configuration of the landscape, the distribution of species and the timing of exposure events turn out to be important determinants of ecosystem effects. The ecological aspects of environmental toxicology will be discussed in Chapter 5.

1.2. DPSIR

On the one hand, environmental toxicology is rooted in more fundamental scientific disciplines like biology and chemistry where curiosity is an important driver for gathering new knowledge. On the other hand, environmental toxicology is a problem-oriented discipline. As such, it is part of the broader field of environmental sciences which analyses the interactions between society and its physical environment in order to promote sustainability. Within this context, knowledge about the interactions of substances with the biotic and abiotic environment is being generated with the ultimate aim to prevent and address potential pollution problems in society. To be able to contribute optimally, an environmental toxicologist should know how pollution problems are structured and what the role of environmental toxicologists is in analysing, preventing and solving such problems. A widely used framework for structuring environmental problems is DPSIR. DPSIR stands for Drivers, Pressures, State, Impacts and Responses (Figure 1). The aim of the current section is to explain the DPSIR framework.

Communication Tool

Communication is essential when analysing and addressing societal issues such as environmental pollution. As an environmental toxicologist, you will have to communicate with fellow scientists to develop a common understanding of the pollution problem, and with policy makers and stakeholders (e.g., producers of chemicals and consumers that are being exposed to chemicals) to explain the scientific state of the art. It is likely that you will use terms like "cause", "source" and "effects". However, not everybody will use and perceive these terms in the same way. Some people may argue that a farmer is the main cause of pesticide pollution, whereas others may argue that it is the pesticide manufacturer, or even the increasing world population. Likewise, some people may perceive the concentration of pesticides in water as an effect of pesticide use, whereas others may refer to the extinction of species when talking about effects. These differences may result in miscommunication, complicating scientific analysis and the search for appropriate solutions.

The DPSIR framework is a tool that helps preventing such communication problems. It provides a common and flexible frame of reference to structure environmental issues by describing these in terms of drivers, pressures, state (variables), impacts and responses (Figure 1). Flexibility is an important characteristic of the framework, enabling adaptation to the problem at hand. The DPSIR framework should not be considered a panacea or used as a mould that rigidly fits all environmental issues. Its main strength is that it stimulates communication between scientists, policy makers and other actors and thereby supports the development of a common understanding.

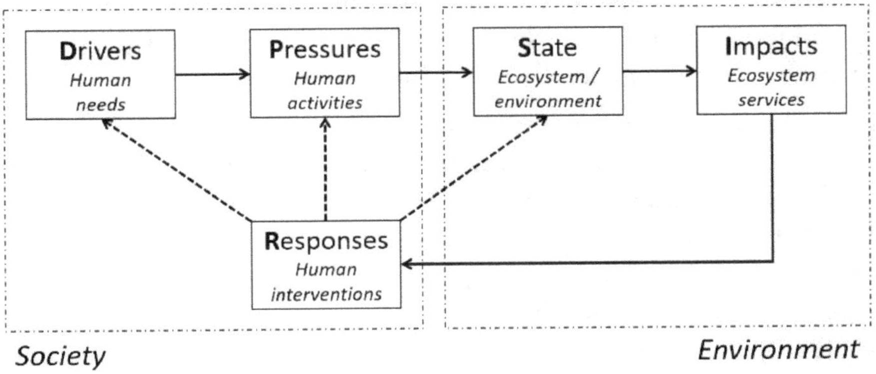

Figure 1.2 The DPSIR framework is a tool to structure environmental issues by organizing the processes in Drivers, Pressures, State (variables), Impacts and Responses. Source: Ad Ragas.

The Framework

The DPSIR framework essentially is a cause-and-effect chain that aims to capture the main processes involved in an environmental issue; from its origin to the changes it triggers in the environment and in society. These processes are organized in five main categories, i.e.:

- are the human needs underlying the human activities that ultimately result in adverse effects. An example is the human need for food resulting in the use of pesticides such as neonicotinoids.
- are human activities initiated to fulfil human needs and resulting in changes in the physical environment that ultimately lead to - often unforeseen – adverse consequences for the environment or certain groups of society that are perceived as problematic, either now or in the future. An example is the use of neonicotinoids in agriculture.
- refers to the status of the physical environment. The state of the environment is often quantified using observable changes in environment parameters, e.g., the concentration of neonicotinoids in water, air, soil and biota.
- are any changes in the physical environment or society that are a consequence of the environmental pressures and that are perceived as problematic by society or some groups in society. An example is the increasing bee mortality that is, at least partly, attributed to the use of neonicotinoids. Or the human health effects of pesticides.
- are all initiatives developed by society to address the issue. These can range from gathering knowledge to developing policy plans and taking measures to mitigate effects or reduce emissions. Examples include the introduction of a risk-based admission procedure for neonicotinoids, the introduction of more efficient spraying techniques, and the development of environmentally friendly pest control techniques.

In principle, any environmental issue can be captured in a DPSIR. But it is important to realize that the labelling of processes as either drivers, pressures, state (variables), impacts or responses is likely to differ between people since the categories are broadly defined and the level of detail in the processes considered may vary. For example, some people may argue that "agriculture" should be classified as a driver, whereas others may argue it is a pressure. Yet other people may deal with this issue by adapting the DPSIR framework, i.e. by adding a new category called "human activities" that is placed in-between the drivers and the pressures. Another typical issue is the labelling of consecutive changes in the physical environment such as rising CO_2 levels, increases in temperature and changes in species abundance. These changes can be labelled as changes in consecutive state variables, i.e. state variables of 1st, 2nd and 3rd order. The idea is that 1st order changes trigger 2nd order changes, e.g. rising CO_2 levels triggering a rise in temperature, and 2nd order changes trigger 3rd order changes, in this case a shift in species abundance. The change in species abundance may also be labelled as an impact, provided this change is perceived as problematic by (groups in) society. The category "impacts" is closely related to the protection goals of risk assessment (see the Section Ecosystem services and protection goals). If there is consensus in society that an impact should be prevented, it becomes a protection goal. All these examples illustrate that the DPSIR framework should be applied in a flexible way and that communication is essential.

Environmental toxicology mainly focuses on the Pressures, State and Impacts blocks of the DPSIR chain. The use of chemicals by society, e.g. in agriculture or in consumer products, and their emission to the environment belongs to the Pressure block. The fate of chemicals in the environment and their accumulation in organisms belongs to the State block. And the adverse effects triggered in ecosystems and humans belong to the Impact block. An important step in risk assessment of chemicals (Chapter 6) is the derivation of safe exposure levels such as the Predicted No Effect Concentration (PNEC) for ecosystems or the Acceptable Daily Intake (ADI) for humans. In terms of DPSIR, this boils down to defining an acceptable impact level (e.g. a zero effect level or a 1 in a million tumor risk) and translating this into a corresponding state parameter (e.g. the chemical concentration in air or water). Fate modelling (Section on Modelling exposure) aims to predict soil, water, air and organisms (all State parameters) based on emission data (a Pressure parameter).

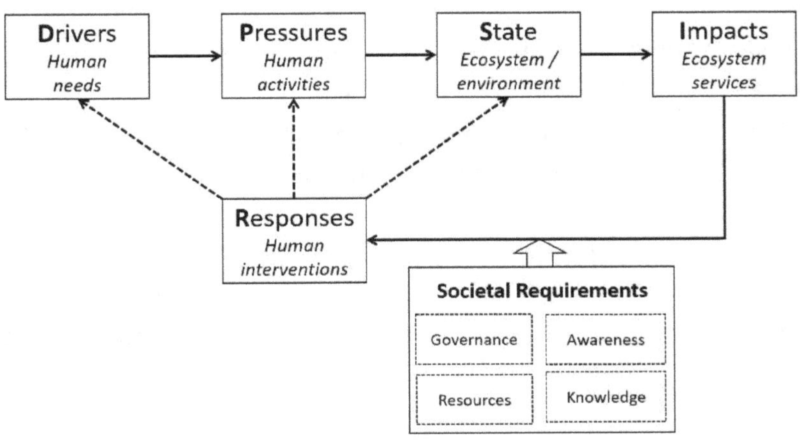

Figure 2. The extended DPSIR framework to put more emphasis on the societal dimension, i.e. governance, awareness, resources and knowledge. Source: Ad Ragas.

The DPSIR framework has been criticized because it tries to capture all processes in cause-and-effect relationships, resulting in a bias towards the physical dimension of environmental issues, e.g. human activities, emissions, physical effects and mitigations measures. The societal dimension is less easily captured, e.g. knowledge generation, governance structures, resources needed to implement measures, awareness and societal framing of the problem (Svarstad *et al.*, 2008). Although the DPSIR framework can been adapted to accommodate some of these aspects (e.g., see Figure 2), it should be acknowledged that it has its limitations. Several alternative frameworks have been developed, and some of these better capture the societal dimension (Gari et al., 2015; Elliott *et al.*, 2017). Nevertheless, DPSIR can be a useful framework to contextualize the problems that are addressed in environmental toxicology. It nicely shows why knowledge on the fate and impact of chemicals (state and impacts) is needed to address pollution issues and that the use of this knowledge is always subject to valuation, i.e. it depends on how society values the adverse effects triggered by the pollution. DPSIR is also widely used by national and international institutes such as the European Environment Agency (EEA), the United States Environmental Protection Agency (US-EPA) and the Organisation for Economic Cooperation and Development (OECD). The DPSIR framework is sometimes also used as a first step in modelling, especially its physical dimension. Once relevant processes have been identified, these are then described quantitatively resulting in models that can be used to predict environmental concentrations or ecological effects of substances based on knowledge about human activities or emissions.

References

Gari, S.R., Newton, A., Icely, J.D. (2015). A review of the application and evolution of the DPSIR framework with an emphasis on coastal social-ecological systems. Ocean & Coastal Management 103, 63-77.

Svarstad, H., Petersen, L.K., Rothman, D., Siepel, H., Wätzold, F. (2008). Discursive biases of the environmental research framework DPSIR. Land Use Policy 25, 116–125.

Elliott, M., Burdon, D., Atkins, J.P., Borja, A., Cormier, R., de Jonge, V.N., Turner, R.K. (2017). "And DPSIR begat DAPSI(W)R(M)!" - A unifying framework for marine environmental management. Marine Pollution Bulletin 118, 27–40.

1.3. SHORT HISTORY

History

From earliest times, man has been confronted with the poisonous properties of certain plants and animals. Poisonous substances are indeed common in nature. People who still live in close contact with nature generally possess an extensive empirical knowledge of poisonous animals and plants. Poisons were, and still are, used by these people for a wide range of applications (catching fish, poisoning arrowheads, in magic rituals and medicines). The first Egyptian medical documentation (written in the Ebers Papyrus) dates from 1550 BC and demonstrates that the ancient Egyptians had an extensive knowledge of the toxic and curative properties of natural products. A good deal is known about the information regarding toxic substances possessed by the Greeks and the Romans. They were very interested in poisons and used them to carry out executions. Socrates, for example, was executed using an extract of hemlock (Conium maculatum). It was also not unusual to use a poison to murder political opponents. Poisons were ideal for that purpose, since it was usually impossible to establish the cause of death by examining the victim. To do so would have required advanced chemical analysis, which was not available at that time.

Early European literature also includes a considerable number of writings on toxins, including the so-called herbals, such as the Dutch "Herbarium of Kruidtboeck" by Petrus Nylandt dating from 1673. Poisoning sometimes assumed the character of a true environmental disaster. One example is poisoning by the fungus Claviceps purpurea, which occurs as a parasite in grain, particularly in rye (spurred rye) and causes the condition known as ergotism. In the past, this type of epidemic has killed thousands of people, who ingested the fungus with their bread. There are detailed accounts of such calamities. For example, in the year 992 an estimated 40,000 people died of ergotism in France and Spain. People were not aware of the fact that death was caused by eating contaminated bread. It was not until much later that it came to be understood that large-scale cultivation of grain involved this kind of risk.

Paracelsus

It was pointed out centuries ago that workers in the mining industry, who came into contact with a variety of metals and other elements, tended to develop specific diseases. The symptoms regularly observed as a result of contact with arsenic and mercury in the mining industry were described in detail by the famous Swiss physician Paracelsus (Figure 1) in his 1567 treatise "Von der Bergsucht und anderen Nergkrankheiten" (miners sickness and other diseases of mining). During the emergence of the scientific renaissance of the 16th century, Paracelsus (1493 - 1541) drew attention to the dose-dependency of the toxic effect of substances. In the words of Paracelsus, "all Ding sind Gifft ... allein die Dosis macht das ein Ding kein Gifft is" (everything is a poison ... it is only the dose that makes it not a poison). This principle is just as valid today. At the same time, it is one of the most neglected principles in the public understanding of toxicology.

A work from the same period "De Re Metallica" by Gergius Agricola (Georg Bauer, 1556), deals with the health aspects of working with metals. Agricola even advised preventive aspects, such as wearing protective clothing (masks) and using ventilation.

Figure 1: A portrait of Paracelsus (PORTRAIT PRESUME DU MEDECIN PARACELSE (1493-1541) (Source: https://commons.wikimedia.org/wiki/File:Paracelsus.jpg)

Scrotum Cancer in Chimney Sweepers: Carcinogenicity of Occupational Exposure

Another example of the rising awareness of the effects of poisons on human health came with the suggestion, by Percival Pott in 1775, that the high frequency of scrotum cancer among British chimney sweepers was due to exposure to soot. He was the first to describe occupational cancer.

> A part of the essay by Percival Pott "The fate of these people seems singularly hard; in their early infancy, they are most frequently treated with great brutality, and almost starved with cold and hunger; they are thrust up narrow, and sometimes hot chimnies, where they are bruised, burned, and almost suffocated; and when they get to puberty, become peculiarly liable to a most noisome, painful, and fatal disease." See the rest of the original text of his essay here.

Soot consists of polycyclic aromatic hydrocarbons (PAHs) and their derivatives. The exposure to soot came with concurrent exposure to a number of carcinogens such as cadmium and chromium. From the 1487 cases of scrotal cancer reported, 6.9 % occurred in chimney sweepers. Scrotal and other skin cancers among chimney sweepers were at the same time also reported from several other countries.

Peppered Moth in Polluted Areas

Changes in the environment due to environmental pollution led to interesting insights in the potential of species to adapt for survival and the role of natural selection in it. A famous example of such micro-evolution is the peppered moth, Biston betularia, that is generally a mottled light color with black speckles. This pattern gives them good camouflage against lichen-covered tree trunks while resting during the day. During the industrial revolution, the massive increase in the burning of coal resulted in the emission of dark smoke turning the light trees in the surrounding areas dark. As a consequence, the dark, melanic form of the peppered moth took over in industrial parts of the United Kingdom during the

1800s. The melanic forms used to be quite rare, but their dark color served as a protective camouflage from bird populations in the polluted areas. This allowed them to become dominant in areas with soot-covered trunks. Two British biologists, Cedric Clarke and Phillip Sheppard, discovered this when they pinned dead moths of the two types on dark and light backgrounds to study their predation by birds. The dark moths had an advantage in the dark forests, a result of natural selection. In areas where air pollution has decreased the melanic form became less abundant again.

Awareness in the 1950s and 1960s

After the second world war, synthetic chemical production became widespread. However, there was limited awareness of the environmental and health risks. In the 1950s, Environmental Toxicology came to light as a result of increasing concern about the impact of toxic chemicals on the environment. This led toxicology to expand from the study of the toxic impacts of chemicals on man to that of toxic impacts on the environment. An important person in raising this awareness was Rachel Carson. Her book "Silent Spring", published in 1962, in which she warned of the dangers of chemical pesticides, triggered widespread public concern on the dangers of improper pesticide use.

DDT is very persistent and tends to concentrate when moving through the food chain. As a consequence, the use of DDT led to very high levels, especially in organisms high in the food chain. Bioaccumulation in birds appeared to cause eggshell thinning and reproductive failure. Because of the increasing evidence of DDT's declining benefits and its environmental and toxicological effects, the United States Department of Agriculture, the federal agency responsible for regulating pesticide use, began regulatory actions in the late 1950s and 1960s to prohibit many of its uses. By the 1980s, the use of DDT was also banned from most Western countries.

Large Environmental Disasters

As a result of large environmental disasters, awareness amongst the general public increased. An enormous industrial pesticide disaster occurred in 1984 in Bhopal, India, when more than 40 ton of the highly toxic methyl isocyanate (MIC) gas leaked from a pesticide plant into the towns located near the plant. Almost 4000 people were killed immediately and 500,000 people were exposed to the poisonous substance causing many additional deaths because of gas-related diseases. The plant was actually initially only allowed to import MIC but was producing it on a large scale by the time of the disaster and safety procedures were far below (international) standards for environmental safety. The disaster made it very clear that this should be changed to avoid other large-scale industrial disasters.

The Sandoz agrochemical spill close to Basel in Switzerland in 1986 was the result of a fire in a storehouse. The emission of large amounts of pesticide caused severe ecological damage to the Rhine river and massive mortality of benthic organisms and fish, particularly eels and salmonids.

At the time of these incidents, environmental standards for chemicals were still largely lacking. The incidents triggered scientists to do more research on the adverse environmental impacts of chemicals. Public pressure to control chemical pollution increased and policy makers introduced instruments to better control the pollution, e.g. environmental permitting, discharge limits and environmental quality standards.

Our Common Future

In 1987, the World Commission on Environment and Development released the report "Our Common Future", also known as the Brundtland Report. This report placed environmental issues firmly on the political agenda, defining sustainable development as "a development that meets the needs of the present without compromising the ability of future generations to meet their own needs". Another influential book was "Our stolen future" written by Theo Colborn and colleagues in 1996. It raised awareness of the endocrine disrupting effects of chemicals released into the environment and threatening (human) reproduction by emphasizing it not only concerns feminization of fish or other organisms in the environment, but also the human species.

SETAC

Before the 1980s, no forum existed for interdisciplinary communication among environmental scientists —biologists, chemists, toxicologists— as well as managers and others interested in environmental issues. The Society of Environmental Toxicology and Chemistry (SETAC) was founded in North America in 1979 to fill this void. In 1991, the European branch started its activities and later SETAC also established branches in other geographical units, like South America, Africa and South-East Asia. SETAC publishes two journals: Environmental Toxicology and Chemistry (ET&C) and Integrated Environmental Assessment and Management (IEAM). SETAC also is active in providing training, e.g. a variety of online courses where you can acquire skills and insights in the latest developments in the field of environmental toxicology. Based on the growth in the society's membership, the meeting attendance and their publications, a forum like SETAC was clearly needed. Read more on SETAC, their publications and how you can get involved here.

Where SETAC focuses on environmental toxicology, international toxicological societies have also been established like EUROTOX in Europe and the Society of Toxicology (SOT) in North America. In addition to SETAC, EUROTOX and SOT, many national toxicological societies and ecotoxicological counterparts or branches became active since the 1970s, showing that environmental toxicology has become a mature field of science. One element indicative of this maturation, also is that the different societies have developed programmes for the certification of toxicologists.

References

Carson, R. (1962). Silent Spring. Houghton Mifflin Company.

Colborn, T., Dumanoski, D., Peterson Myers, J. (1996). Our Stolen Future: Are We Threatening Our Fertility, Intelligence, and Survival? A Scientific Detective Story. New York: Dutton. 306 p.

World Commission on Environment and Development (1987). Our Common Future. Oxford: Oxford University Press. p.27.

2 Chapter ENVIRONMENTAL CHEMISTRY, CHEMICALS

2.1. INTRODUCTION

Environmental toxicology deals with the negative effects of the exposure to chemicals we regard as pollutants (or contaminants/toxicants). Environmental toxicants receive a lot of media attention, but many critical details are getting lost or are easily forgotten. The clip "You, Me, DDT" shows the discovery of the grandson of the works of his grandfather, the Swiss inventor of the insecticide DDT, Paul Hermann Müller, who received the Nobel Prize for Medicine for that in 1948 (see also Section 1.3). The clip "Stop the POPs" interviews (seemingly) common people about one of the most heavily regulated group of pollutants.

Organisms, including humans, have always been exposed to chemicals in the environment and rely on many of these chemicals as nutrients. Volcanoes, flooding of acid sulfur lakes, and forest fires have caused widespread contamination episodes. Organisms are also in many cases directly or indirectly involved in the fate and distribution of undesirable chemicals in the environment. Many naturally occurring chemicals are toxicants already (see also Section 1.3), think for example about:

- local arsenic or mercury hotspots in the Earth's crust, contaminating water pumps or rice irrigation fields;
- plant-based defence chemicals such as alkaloids, morphine in poppy seeds, juglone from black walnut trees);
- fungal toxins, such as mycotoxins threatening grain storage depots after harvests;
- bacterial toxins, such as the botulinum toxin, a neurotoxic protein produced by the bacterium Clostridium botulinum which is the most acutely lethal toxin known at ~10 ng/kg body weight when inhaled;

- phycotoxins, produced by algae, in mass algal blooms or those that may end up at dangerous levels in shell food;
- zootoxins in animals, such as venom of snakes and defensive toxins on the skin of amphibians.

Human activities have had an enormous impact on the increased exposure to natural chemicals as a result of, for example, the mining and use of metals, salts and fossil fuels from geological resources. This is for example the case for many metals, nutrients such as nitrate, and organic chemicals present in fossil fuels. Additionally, the industrial synthesis and use of organic chemicals, and the disposal of wastes, have resulted in a wide variety of hazardous chemicals that had either never existed before, or at least not in the levels or chemical form that occur nowadays in our heavily polluted global system. These are typically organic chemicals that are referred to as anthropogenic ('due to humans in nature') or xenobiotic ('foreign to organisms') chemicals. In this chapter we aim to clarify the key properties and functionalities of the most common groups of pollutants as a result from human activities, and provide some background on how we can group them and understand their behaviour in the environment.

In the field of environmental toxicology, we are most often concerned about the effects of two distinct types of contaminants: metals and organic chemicals. In some cases other chemicals, such as radioactive elements, may also be important while we could also consider the ecological effects of highly elevated nutrient concentrations (eutrophication) as a form of environmental toxicology.

Metals

Metals and metalloids (elements intermediate between metals and non-metals) comprise the majority of the known elements. They are mined from minerals and used in an enormous variety of applications either in their elemental form or as chemicals with inorganic or organic elements or ions. Many metals occur as cations, but many processes influence the dissolved form of metals. Aluminium for example for example is only present under very acidic conditions as dissolved cation (Al^{3+}), while at neutral pH the metal speciates into for example certain hydroxides ($Al(OH_3)^0$). Mercury as free ion is present at (Hg^{2+}) but due to microbial transformation the highly toxic product methylmercury (CH_3Hg^+, or $MeHg^+$) is formed. Mining and processing of metals together with disposal of metal-containing wastes are the main contributors to metal pollution although sometimes metals are introduced deliberately into the environment as biocides. The widely used pesticide copper sulfate in e.g. grape districts is an example (LINK on comparison to glyphosate here). More information on metals considered to be environmental pollutants is given in Section 2.2.1.

Organic Chemicals

Organic chemicals are manufactured to be used in a wide variety of applications. These range from chemicals used as pesticides to industrial intermediates, fossil fuel related hydrocarbons, additives used to treat textiles and polymers, such as flame retardants and plasticisers, and household chemicals such as detergents, pharmaceuticals and cosmetics.

Organic chemicals that we regard as environmental pollutants include a huge variety of different structures and have a wide variety of properties that influence environmental distribution and toxicity. With such a wide variety of chemicals to deal with, it is useful to classify them into groups. Depending on our interest, we can base this classification on different aspects, for example on their chemical structure, their physical and chemical properties, the applications the chemicals are used in, or their effects on biological systems. These aspects are of course closely related to their chemical structure as this is the basis of the properties and effects of chemicals. An overview of different ways of classifying environmental contaminant (sometimes referred to as ecotoxicants) is shown in Tables 1A, 1B, and 1C.

Table 1A. Grouping options of organic contaminants with specific chemical structures

Term	Characteristics	Examples
Hydrocarbons	More CHx units: higher hydrophobicity/ lipophilicity, and lower aqueous solubility	hexane
Polycyclic aromatic hydrocarbons	Combustion products. Flat structure	naphthalene, B[a]P
Halogenated hydrocarbons	H substituted by fluor, chlorine, bromide, iodine. Often relatively persistent	PCB, DDT, PBDE
Dioxins and furans	Combustion/industrial products, one or two oxygen atoms between two aromatic rings. Highly toxic.	TCDD, TCDF
Organometallics	Organic chemicals containing metals, used e.g. in anti-fouling paints	tributyltin
Organophosphate pesticides	Phosphate esters, often connecting two lipophilic groups. Act on nervous system	chlorpyrifos
Pyrethroids	Usually synthetic pesticides based on natural pyrethrum extracts	fenvalerate
Neonicotinoids	Synthetic insecticides with aromatic nitrogen, related to the alkaloid nicotine	imidacloprid
... Endless varieties / combinations and ...	...too many characteristics to list	...

Table 1B. Grouping options of organic contaminants with specific properties

Term	Characteristics	Examples
Hydrocarbons	More CHx units: higher hydrophobicity/ lipophilicity, and lower aqueous solubility	hexane
Polycyclic aromatic hydrocarbons	Combustion products. Flat structure	naphthalene, B[a]P
Halogenated hydrocarbons	H substituted by fluor, chlorine, bromide, iodine. Often relatively persistent	PCB, DDT, PBDE
Dioxins and furans	Combustion/industrial products, one or two oxygen atoms between two aromatic rings. Highly toxic.	TCDD, TCDF
Organometallics	Organic chemicals containing metals, used e.g. in anti-fouling paints	tributyltin
Organophosphate pesticides	Phosphate esters, often connecting two lipophilic groups. Act on nervous system	chlorpyrifos
Pyrethroids	Usually synthetic pesticides based on natural pyrethrum extracts	fenvalerate
Neonicotinoids	Synthetic insecticides with aromatic nitrogen, related to the alkaloid nicotine	imidacloprid
... Endless varieties / combinations and ...	...too many characteristics to list	...

Table 1C. Grouping options of organic contaminants with specific usage

Term	Characteristics	Examples
Pesticides Herbicides Insecticides Fungicides Rodenticides Biocides	Toxic to pests Toxic to plants Toxic to insects Toxic to fungi Toxic to rodents Toxic to many species	DDT atrazine, glyphosate Chlorpyrifos, parathion Phenyl mercury acetate Hydrogen cyanide Benzalkonium
Pharmaceuticals	Specifically bioactive chemicals with often (un) known side effects. Many bases.	diclofenac (pain killer), iodixanol (radio-contrast), carbemazepine, prozac
Drugs of abuse	Often opioid based but also synthetic designer drugs with similar activity. Many are are ionogenic bases.	cannabinoids, opioids, amphetamine, LSD
Veterinary Pharmaceuticals	Can include relatively complex (ionogenic) structures	antibiotics, antifungals, steroids, non-steroidal anti-inflammatories

Term	Characteristics	Examples
Industrial Chemicals	Produced in large volumes by chemical industry for a wide array of products and processes	phenol
Fuel products	Flammable chemicals	kerosene
Refrigerants and propellants	Small chemicals with specific boiling points	freon-22
Cosmetics/personal care products	Wide varieties of specific ingredients of formulations that render specific properties of a product	sunscreen, parabenes
Detergents and surfactants	Long hydrophobic hydrocarbon tails and polar/ ionic headgroups	Sodium lauryl sulfate (SLS), benzalkonium
Food and Feed Additives	To preserve flavor or enhance its taste, appearance, or other qualities	"E-numbers", acetic acid = E260 in EU, additive 260 in other countries

Chapter 2 mostly discusses groups of chemicals in separate modules according to the specific environmental properties in Table 1B (Section 2.2) and specific applications in Table 1C (Section 2.3), according to which certain regulations apply in most cases. The property classifications can be based on (often interrelated) properties such as solubility (in water), hydrophobicity (tendency to leave the water), surface activity (tendency to accumulate at surfaces of two phases, such as for "surfactants"), polarity, neutral or ionic chemicals and reactivity. Other classifications very important for environmental toxicology are based on environmental behaviour or effects, such as persistency ("P" increasing problems with increased emissions), bioaccumulation potential ("B", up-concentration in food chains), or type of specific toxic effects ("T"). The influence of specific chemical structures such as in Table 1A is further clarified in the current introductory chapter in order to better understand the basic chemical terminology.

Structures of Organic Chemicals and Functional Groups

◉ Hydrocarbons and polycyclic aromatic hydrocarbons

As the name suggest, hydrocarbons contain only carbon and hydrogen atoms and can therefore be considered to be the simplest group of organic molecules. Nevertheless, this group covers a wide variety of aliphatic, cycloaliphatic and aromatic structures (see Figure 1 for some examples) and also a wide range of properties. What this group shares is a low solubility in water with the larger molecules being extremely insoluble and accumulating strongly in organic media such as soil organic matter.

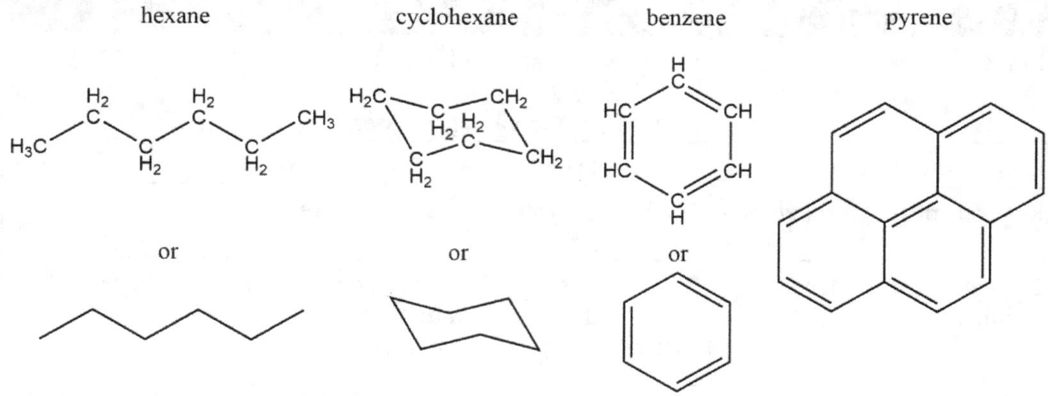

Figure 1. Examples of hydrocarbons.

As a result of the ability of carbon to form strong bonds with itself and other atoms to form structures containing long chains or rings of carbon atoms there is a huge and increasing number (millions) of organic chemicals known. Chemicals containing only carbon and hydrogen are known as hydrocarbons. Aliphatic molecules consist of chains of carbon atoms as either straight or branched chains. Molecules containing multiple carbon-carbon bonds (C=C) are known as unsaturated molecules and can be converted to saturated molecules by addition of hydrogen.

Cyclic alkanes consist of rings or carbon atoms. These may also be unsaturated and a special class of these is known as aromatic hydrocarbons, for example benzene in Figure 1. The specific electronic structure in aromatic molecules such as benzene makes them much more stable than other hydrocarbons.

Multiple aromatic rings linked together make perfectly flat molecules, such as pyrene in Figure 1, that can be polarized to some extent because of the shared electron rings. In larger sheets, these polycyclic aromatic molecules also make up the basic graphite structure in pencils, and also typically represent the strongly adsorbing surfaces of black carbon phases such as soot and activated carbon.

The structures of organic chemicals help to determine their properties as behaviour in the environment. At least as important in this regard, however, is the presence of functional groups. These are additional atoms or chemical groups that are present in the molecule that have characteristic chemical effects such as increasing or decreasing solubility in water, giving the chemical acidic or basic properties or other forms of chemical reactivity. The common functional groups are shown in Table 2.

Table 2. Common Functional Groups, where R are carbon backbone or hydrogen units

Class	Functional group	General Formula	Examples
Halohydrocarbons	—X (F, Cl, Br, I)	R—X	CH_3I (iodomethane)
Alcohols	—OH	R—OH	CH_3OH (methanol)
Ethers	—O—	R—O—R'	CH_3OCH_3 (dimethylether)
Aldehydes	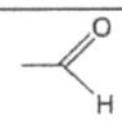	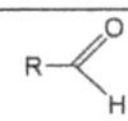	CH_2O (methanal)
Ketones	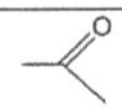	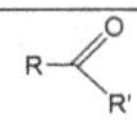	CH_3COCH_3 (propanone, or acetone)
Carboxylic Acids	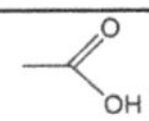	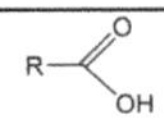	CH_3COOH (ethanoic acid, or acetic acid)
Esters	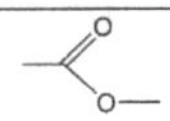	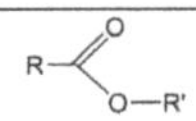	$CH_3COOCH_2CH_3$ (ethyl acetate)
Amines	$—NH_2$	R—NR'R"	CH_3NH_2 (methylamine, or aminomethane)
Amide	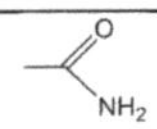	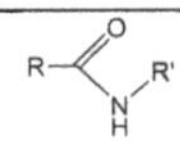	$CH_3CON(CH_3)_2$ (N,N-dimethylacetamide)
Sulphonamide	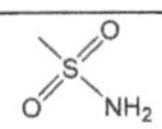	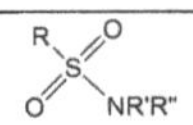	CH_3SOONH_2 (methylsulfonamide)
Carbamate	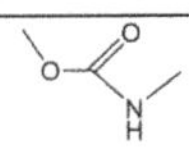	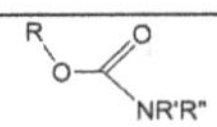	X
Nitro	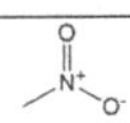	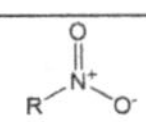	CH_3NOO (nitromethane)

◉ Halogenated hydrocarbons: first generation pesticides

The first organic chemical recognised as an environmental pollutant was the insecticide DDT (see clip 1 at the start of this chapter). It later became clear that other organochlorine pesticides such as lindane and dieldrin (Table 3) were also widely distributed in the environment. This was also the case for polychlorinated biphenyls (PCBs) and other organochlorinated industrial chemicals. These chemicals all share a number of undesirable properties such as environmental persistence, very low solubility in water and high level of accumulation in biota to potentially toxic levels. Many organochlorines can be viewed as hydrocarbons in which hydrogen atoms have been replaced by chlorine. This makes them even less soluble than the corresponding hydrocarbon due to the large size of chlorine atoms. In addition, chlorination also makes the molecules more chemically stable and therefore contributes to their environmental persistence. Other organochlorines contain additional functional groups, such as the ether bridges in PCDDs and PCDFs (better known as dioxins and dibenzofurans) and ester groups in the 2,4-D and 2,4,5-T herbicides. Many organochlorines were applied very successfully in huge quantities as pesticides for decades before their negative effects such as persistence and accumulation in biota became apparent. It is therefore no coincidence that the initial set of Persistent Organic Pollutants (POPs) identified in the Stockholm Treaty (see below) as chemicals that should be banned were all organochlorines, as shown in Table 3.

Table 3. Key persistent organic pollutants, also named POPs – the Dirty Dozen

Chemical	Use	Structure
Aldrin/ Dieldrin/ Endrin	crop insecticides (corn, cotton, grains)	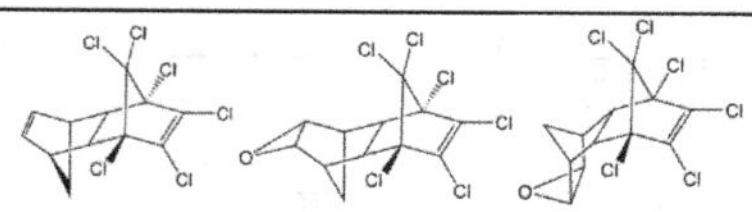
Chlordane/ Heptachlor	insecticides (potatoes, citrus vegetables, cotton/termites, soil insects)	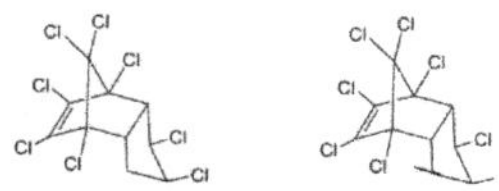
DDT/ Hexachloro -benzene	crop insecticide (cotton) / fungicide for seed treatment	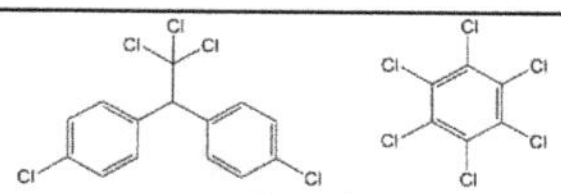
Mirex/ Toxaphene	insecticide (termites, fire ants/ livestock and crops)	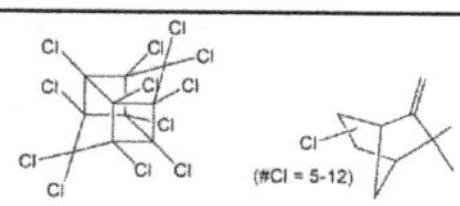 (#Cl = 5-12)
PCBs	industrial (heat exchange fluid for e.g. electrical transformers, paint and plastic additive)	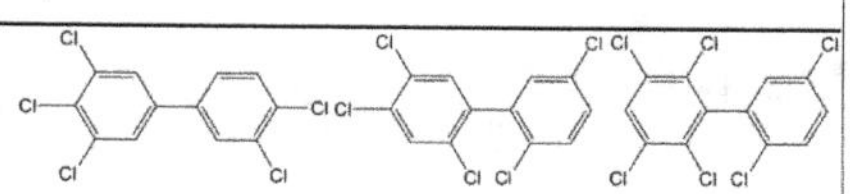 PCB 81 (co-planar structure); PCB 101, PCB 151 (non-planar structures)
Dioxins/ Furans	unintentionally produced during combustion	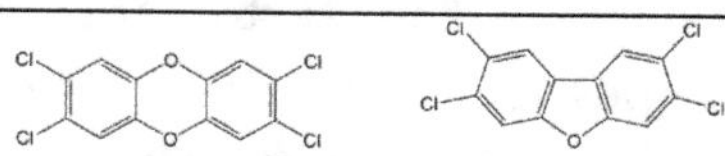 TCDD (2,3,7,8 tetrachlorodioxin); TCDF (2,3,7,8 tetrachlorofuran)

As well as chlorine, other halogens such as bromine and fluorine are used in important groups of environmental contaminants. Organobromines are best known as flame retardants and have been applied in large quantities to improve the fire safety of plastics and textiles. They share many of the same undesirable properties of organochlorines and several classes have now been taken out of production. Organofluorines are another important class of halogenated chemicals, and part of the well-known group of ozone depleting CFCs (Section 2.3.6). In particular, per-and polyfluoralkyl substances are widely used as fire-stable surfactants in fire-fighting foams, as grease and water resistant coatings and in the production of fluoropolymers such as Teflon. Organofluorines are much more water soluble and much less bioaccumulative than organochlorines and organobromines but are extremely persistent in the environment.

The recognition of these organochlorines as harmful environmental contaminants eventually resulted in measures to restrict their manufacture and use in the Stockholm Convention on Persistent Organic Pollutants signed in 2001 to eliminate or restrict the production and use of persistent organic pollutants (POPs). This initial list of POPs has been subsequently augmented with other harmful halogenated organic pollutants up to a total of 29 chemicals, which are either to be eliminated, restricted, or required measured to reduce unintentional releases. POPs are further discussed in section 2.2.4.

Additional POPs to eliminate include: chlordecone, lindane (hexachlorocyclohexane), pentachlorobenzene, endosulfan, chlorinated naphthalenes, hexachlorobutadiene,tetrabromodiphenylether, and pentabromodiphenylether decabromodiphenyl ether (BDEs)

◉ Alternatives for the organochlorine pesticides: effective functional groups

Since the signing of the Stockholm Convention, organochlorine pesticides have been replaced in most countries by more modern pesticide types such as the organophosphorus and carbamate insecticides. These compounds are less persistent in the environment, but still could pose elevated risks to environments surrounding the agricultural sites, and increased levels on food produced on these agricultural sites. The very toxic organophosphorus neurotoxicant parathion has been in use since the 1940s, and has the typical two lipophilic side chains on two esters (ethyl units), as well as a polar unit. Parathion has caused hundreds of fatal and non-fatal intoxications worldwide and as a result it is banned or restricted in 23 countries. The relatively comparable organophosphate structure of diazinon has been widely used for general-purpose gardening and indoor pest control since the 1970s, but residential use was banned in the U.S. in 2004. In Californian agriculture however, 35000 kg diazinon was used in 2012. The carbamate based insecticide carbaryl is toxic to target insects, and also non-target insects such as bees, but is detoxified and eliminated rapidly in vertebrates, and not secreted in milk. Although illegal in 7 countries, carbaryl is the third-most-used insecticide in the U.S., approved for more than 100 crops. In 2012, 52000 kg carbaryl was used in California, while this was 3 times more in 2000. Neonicotinoid insecticides, with the typical aromatic ring containing nitrogen, form a third generation of pesticide structures. Imidacloprid is currently the most widely used insecticide worldwide, but as of 2018 banned in the EU, along with two other neonicotinoids clothianidin and thiamethoxam.

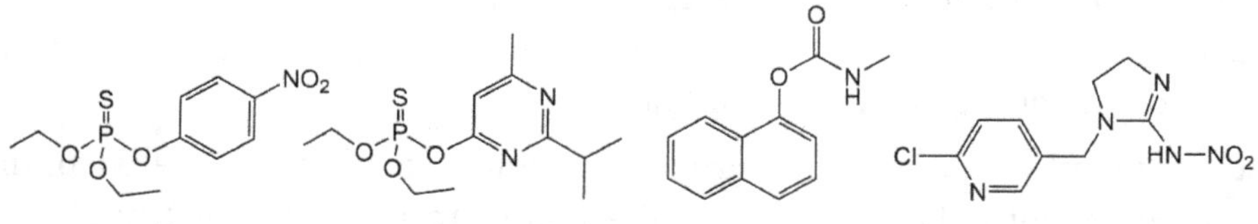

parathion diazinon carbaryl imidacloprid

Figure 2. Some examples of second and third generation replacements of organochlorine pesticides.

- Relatively simple and (very) complex pollutants

As well as the pesticides discussed above, many other chemicals are brought into the environment inadvertently during their manufacture, distribution and use and the range of chemicals recognised as problematic environmental contaminants has expanded enormously. These include fossil fuel-related hydrocarbons, surfactants, pigments, biocides and chemicals used as pharmaceuticals and personal care products (PPCPs). Figure 3 gives an illustrative overview of the major routes by which PPCPs, but also many other anthropogenic contaminants other than pesticides, are released into the environment. Particularly wastewater treatment systems form the main entry point for many industrial and household products.

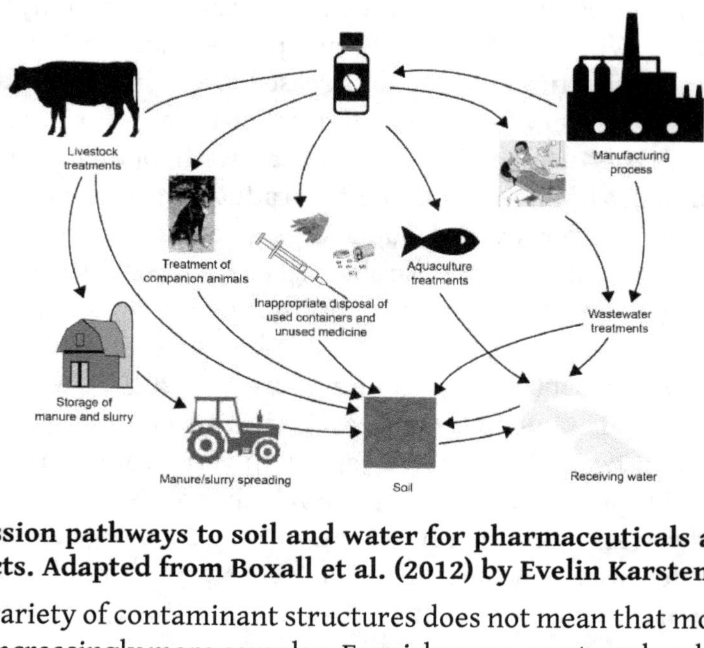

Figure 3. Emission pathways to soil and water for pharmaceuticals and personal care products. Adapted from Boxall et al. (2012) by Evelin Karsten-Meessen.

The wide variety of contaminant structures does not mean that most chemicals have become increasingly more complex. For risk assessment, molecular properties such as water solubility, volatility and lipophilicity are often estimated based on quantitative structure-property relationships (Section 3.4.3). With increasingly complex structures, such property-estimations based on the molecular structure become more uncertain.

The antibiotic erythromycin for example (Figure 4), is a very complex chemical structure ($C_{37}H_{67}NO_{13}$) that has 13 functional units along with a 14 member ring. In addition, the tertiary nitrogen group is an amine base group that can give the molecule a positive charge upon protonation, depending on the environmental pH. Erythromycin is on the World Health Organization's List of Essential Medicines (the most effective and safe medicines needed in a health system), and therefore widely used. Continuous emissions in waste streams pose a potential threat to many ecosystems, but many environmentally and toxicologically relevant properties are scarcely studied, and poorly estimated.

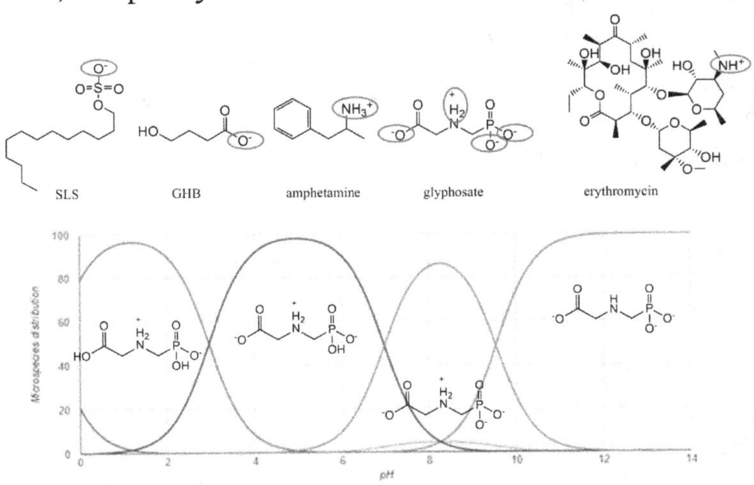

Figure 4. Relatively simple or (very) complex chemicals? Glyphosate speciation profile (chemicalize.org) for the 4 dominant glyphosate different species. Glyphosate in soil (pH4-8) predominantly occurs with 3 charged groups (net charge -1), partly with 4 charged groups (-2). The soap SLS is always negatively charged, GHB predominantly negative (pK_a 4.7), amphetamine predominantly positive (pK_a 9.9), erythromycin predominantly positive (pK_a 8.9)

There are also many contaminants or toxicants with a seemingly simple structure. Many surfactants are simple linear long chain hydrocarbons with a polar or charged headgroup (Figure 4). The illicit drug amphetamine has only a benzene ring and an amine unit, the illicit drug GHB only an alcohol and a carboxylic acid, the herbicide glyphosate only 16 atoms. Still, these 4 chemical examples also have acidic or basic units that often result in predominantly charged organic molecules, which also strongly influences their environmental and toxicological behaviour (see sections on PMOCs and Ionogenic Organic Compounds). In case of glyphosate, the chemical has 4 differently charged forms depending on the pH of the environment. At common pH of 7-9, glyphosate has all charged groups predominantly ionized, making it very difficult to derive calculations on environmental properties.

References

Boxall, A. B. A., Rudd, M. A., Brooks, B. W., Caldwell, D. J., Choi, K., Hickmann, S., Innes, E., Ostapyk, K., Staveley, J. P., Verslycke, T., Ankley, G. T., Beazley, K. F., Belanger, S. E., Berninger, J. P., Carriquiriborde, P., Coors, A., DeLeo, P. C., Dyer, S. D., Ericson, J. F., Gagné, F., Giesy, J. P., Gouin, T., Hallstrom, L., Karlsson, M. V., Larsson, D. G. J., Lazorchak, J. M., Mastrocco, F., McLaughlin, A., McMaster, M. E., Meyerhoff, R. D., Moore, R., Parrott, J. L., Snape, J. R., Murray-Smith, R., Servos, M. R., Sibley, P. K., Oliver Straub, J., Szabo, N. D., Topp, E., Tetreault, G. R., Trudeau, V. L., Van der Kraak, G. (2012). Pharmaceuticals and personal care products in the environment: what are the big questions? Environmental Health Perspectives 120, 1221-1229.

2.2 POLLUTANTS WITH SPECIFIC PROPERTIES

2.2.1. Metals and metalloids

Introduction

The majority of the elements in the periodic table consists of metals: Figure 1.

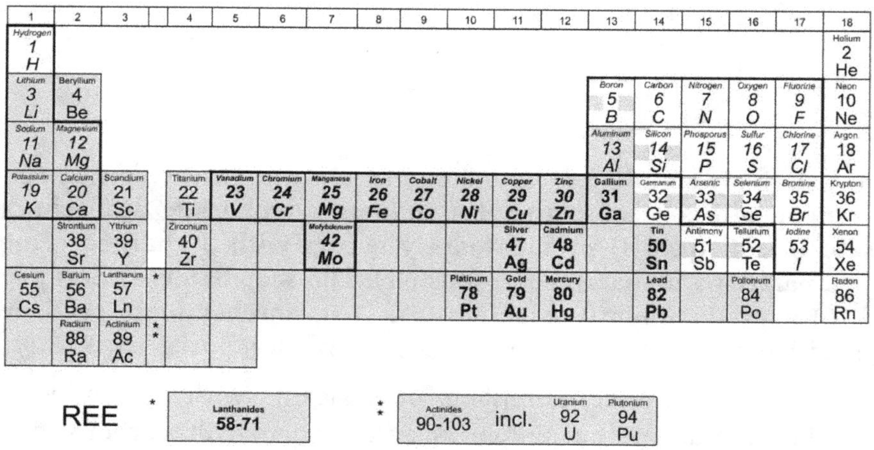

Figure 1. Periodic table of elements, with the most important elements for Environmental Toxicology shown. The shaded elements are metals, the partially shaded elements are metalloids. Bold lettered metals area heavy metals (specific density > 5 g/cm3). Elements shown within bold lines (and in italics) are essential elements. The Lanthanides and Actinides together are the rare earth elements (REEs). (Source: Steven Droge).

The distinction between metals and heavy metals (density relative to water < or >5 g cm-3) is not very meaningful for such a heterogeneous group of elements with rather different biological and chemical properties. The rare earth elements (REEs), lanthanides and actinides, have, for example, a high density or specific weight but are usually not considered heavy metals because of their rather different chemical behaviour. Metalloids have both metallic and non-metallic properties or are nonmetallic elements that can combine with a metal to produce an alloy. Figure 1 shows the periodic table of elements, indicating the groups of (heavy) metals, metalloids and rare earth elements. Also indicated in Figure 1 are the elements that are known to be essential to life and include besides C, H, O and N, the major essential elements Ca, P, K, Mg, Na, Cl and S, the trace elements Fe, I, Cu, Mn, Zn, Co, Mo, Se, Cr, Ni, V, Si, As and B (the latter only for plants) and some elements that may support physiological functions at ultra-trace levels (Li, Al, F and Sn) (Walker *et al.*, 2012).

Chemical and Physical Properties

Except for mercury, most pure metals are solid at room temperature. In general, metals are good electrical and thermal conductors having high luster and malleability. Upon heating, metals readily emit electrons. These descriptors of metals, however, are not very helpful when having to deal with elements that do not exist prominently in the pure elemental state, but rather are present as metal compounds, complexes, and ions at fairly low environmental concentrations.

More useful are characteristics that influence metal transport between environmental compartments and their interaction with abiotic and biotic components of the environment. The speciation, the chemical form in which an element occurs (e.g., oxidized, free ion or complexed to inorganic or organic molecules), determines its transport and interaction in the environment (see Section on Metal Speciation). Chemical bonding is determined by outer orbital electron behavior, with metals tending to lose electrons when reacting with nonmetals. In many normal biological reactions, metals are cofactors within coenzymes (e.g. in vitamins) and can act as electron acceptors and donors during oxidation and reduction reactions (Newman, 2015).

Nieboer and Richardson (1980) proposed a classification, based on the equilibrium constant for the formation of metal complexes. They distinguished:

Class A-metals: acting as hard Lewis acids (electron acceptors) with high affinity for oxygen-containing groups in macromolecules, such as carboxyl and alcohol groups. Al, Ba, Be, Ca, K, Li, Mg, Na and Sr belong to this group;

Class B-metals: acting as soft Lewis acids with high affinity for nitrogen- and sulphur-containing groups in macromolecules, such as amino and sulphydryl groups. This group includes Ag, Au, Bi, Hg, Pd, Pt and Tl.

In addition, an intermediate or borderline group is defined, in which the type A or B characteristics are less pronounced. As, Cd, Co, Cr, Cu, Fe, Mn, Ni, Pb, Sb, Sn, Ti, V, and Zn belong to this group.

This classification of metals is highly relevant for the transport across cell membranes, the intercellular storage in granules and the induction of metal-binding proteins as well as for their behaviour in the environment in general.

Occurrence

(Heavy) metals and rare earth elements are diffusely distributed over the Earth, but at some places certain elemental combinations are highly concentrated (in metal ores). Despite this diffuse distribution, differences in background metal concentrations in soils can be large, depending on the type and origin of rock or sediment (Table 1).

Table 1. Background concentrations (mg/kg dry weight) of (heavy) metals and metalloids in crust material and median and maximum concentrations in different top soils across the world. Derived from Kabata-Pendias and Mukherjee (2007) and Alloway (2013).

Element	Upper crust	Median	Maximum
Ag	0.06-0.07	0.05-1.0	3.15
As	1.8-2.0	1.9-7.3	123
Be	3	0.92-1.9	
Cd	0.1	0.13-0.7	41
Co	10-12	2.4-18	347
Cr	35-100	9.9-96	5,910
Cu	14-55	7-109	3,088
Hg	0.07	0.037-0.086	3.5
La	30	23-37	
Li	20	13-28	
Mn	527-900	255-752	42,603
Mo	1.4-1.5	<0.5-3.0	74
Ni	19-20	4.9-50	4,955
Pb	17	8.4-24.8	2,635
Se	0.05	0.14-0.74	17
Sn	2.5	<1.0-1.68	110
Ti	4,400	2,900-15,480	
Tl	0.5	0.23-0.66	24
V	53-135	18-320	490
Zn	52-70	26.8-89	6,173
Zr	165	92-421	

In general, volcanic rock (e.g. basalt) contains high and sedimented rock (e.g. limestone) low metal levels. But there is no relation between metal concentrations in the Earth's crust and the elemental requirements of organisms.

Emissions of Metals

Upon weathering of stone formations and ores, elements are released and enter local, regional and global biogeochemical cycles. Depending on their water solubility and on soil properties and vegetation, metals may be transported through the environment and deposited or precipitated at places close to or far away from their source.

Volcanoes take account of the largest natural input of metals to the environment but the concentrations of these metals in the soil are rarely elevated to toxic levels due to the massive dilution which takes place in the atmosphere. Permanently active volcanoes may be an important local source of (metal) pollution.

A special case is arsenic, that may occur as a natural element of soils. At some places, As levels are fairly high, particularly in ground water. High As-groundwater areas are found in Argentina, Chile, Mexico, China and Hungary, and also in Bangladesh, India (West Bengal), Cambodia, Laos and Vietnam. In the latter countries, especially in the Bengal Basin, millions of wells have been dug to provide safe drinking water. Irrigation pumping leads to an inflow of oxygen and organic carbon, which causes a mobilisation of arsenic normally bound to ferric oxyhydroxides in these soils. As a result in many wells dissolved As concentrations are exceeding the World Health Organisation (WHO) guideline value of 10 μg/L for drinking water.

Important anthropogenic sources of metals in the environment include:

- Metal mining, which may also lead to an enormous physical disturbance of the environment (destruction of ecosystems).
- Metal smelting.
- Use of metals in domestic and industrial products, and the discharge of domestic waste and sewage.
- Metal-containing pesticides, e.g. 'Bordeaux Mixture (copper sulphate with lime ($Ca(OH)_2$), used as a fungicide in viniculture, hop-growing and fruit-culture, and metal-containing fungicides, such as organo-tin compounds.
- The use of metals and especially of REEs in microelectronics.
- Energy-producing industries burning coal and oil, and producing metal-containing fly ash.
- Transport of energy and traffic making use of electricity, giving rise to corrosion of electric wires and pylons.

- Non-metal industries, e.g. leather (chromium) and cement production (thallium).
- Using Tetra Ethyl Lead (TEL) as anti-knocking agent in petrol (nowadays banned in most countries) and the use of catalysts in cars (platinum, palladium).

Anthropogenic releases of many metals, such as Pb, Zn, Cd and Cu, are estimated to be between one and three orders of magnitude higher than natural fluxes (Depledge et al. 1998). An estimated amount of up to 50,000 tonnes of mercury are released naturally per year as a result of degassing from the Earth's crust, but human activities account for even larger emissions (Walker et al. 2012).

References

Alloway, B.J. (2013). Heavy Metals in Soils. Trace Metals and Metalloids in Soils and their Bioavailability. Third Edition. Environmental Pollution, Volume 22, Springer, Dordrecht.

Depledge, M.H., Weeks, J.M., Bjerregaard, P. (1998). Heavy metals. In: Calow, P. (Ed.). Handbook of Ecotoxicology. Blackwell Science, Oxford, pp. 543-569.

Kabata-Pendias, A., Mukherjee, A.B. (2007). Trace Elements from Soil to Human. Springer Verlag, Berlin.

Newman, M.C. (2015). Fundamentals of Ecotoxicology. The Science of Pollution. Fourth Edition. CRC Press, Taylor & Francis Group. Boca Raton.

Nieboer, E., Richardson, D.H.S. (1990). The replacement of the nodescript term 'Heavy metals' by a biologically and chemically significant classification of metal ions. Environmental Pollution (Ser. B) 1, 3-26.

Walker, C.H., Hopkin, S.P., Sibly, R.M., Peakall, D.B. (2012). Principles of Ecotoxicology, Fourth Edition. CRC Press Taylor & Francis Group, London.

2.2.2. Radioactive Compounds

Introduction

Naturally occurring radionuclides are omnipresent in the environment and exposure to radiation is unequivocally related to life on Earth. Every day we are exposed to cosmic radiation, radon exhalation from the soil and radioactive potassium naturally present in our bodies. Moreover, radionuclides and ionising radiation are successfully applied in many domains such as nuclear medicine, research applications, energy production, food preservation and other industrial activities. To be able to positively apply radionuclides or ionising radiation and to evaluate the impact on man and environment in case of a contamination scenario, it is important to understand the process of radioactive decay, the different types of radiation and radionuclides and how ionising radiation interacts with matter.

Radioactive Decay

Radioactivity is the phenomenon of spontaneous disintegration or decay of unstable atomic nuclei to form energetically more stable ones (Krane, 1988). Within this process, particles (e.g. protons, neutrons) and/or radiation (photons) can be emitted. This radioactive decay is irreversible and after one or more transformations, a stable, non-radioactive atom is formed.

Radioactive decay is considered a stochastic phenomenon as it is impossible to predict when any given atom will disintegrate. However, the probability per unit time that an unstable nucleus can decay is described by the disintegration or decay constant λ [s-1]. The fact that this probability is constant, forms the basic assumption of the statistical theory of radioactive decay.

Radioactive decay follows an exponential function (Eq. 1, Figure 1) with N0 the number of nuclei at time 0, N(t) the remaining nuclei at time t and λ the decay constant [s-1].

2.2.3. Industrial Chemicals

Introduction

The chemical industry produces a wide variety of chemicals that find use in industrial process and as ingredients in day-to-day products for consumers. Instead of chemicals, 'substances' may be a more carefully worded description as it also includes complex mixtures, polymers and nanoparticles. Many substances are produced by globally distributed companies in very high volumes, ranging for example from 100 - 10,000 tonnes (1 tonne = 1000 kg) per year. Worldwide, governments have tried to control and assess chemical safety, as nicely summarized on the ChemHAT website. Australia for example, has the Industrial Chemicals (*Notification and Assessment*) Act 1989 (2013 version). Just like elsewhere in the world, in the European Union (EU) a variety of regulatory institutes at all levels of government used to perform safety assessments regarding the use of substances in products, and how these are emitted into waste streams. This changed dramatically in 2007.

On June 1st, 2007 (Figure 1), a new EU regulation went into force called REACH (official legislation documents C 1907/2006; about REACH; EU info on REACH). This law reversed the role of governments in chemical safety assessment, because it placed the burden of proof on companies that manufacture a chemical, import a chemical into the EU, or apply chemicals in their products. Within REACH companies must identify and manage the risks linked to the chemicals they manufacture and market in the EU. REACH stands for Registration, Evaluation, Authorisation and Restriction of Chemicals. China soon followed with the analogous "China REACH" in 2010, and then came South Korea in 2015 with "K-REACH". The main focus in this module is on EU-REACH as the leading and well documented example. Other legislation regulating industrial chemicals can often be easily found online, e.g. via the ChemHAT link above.

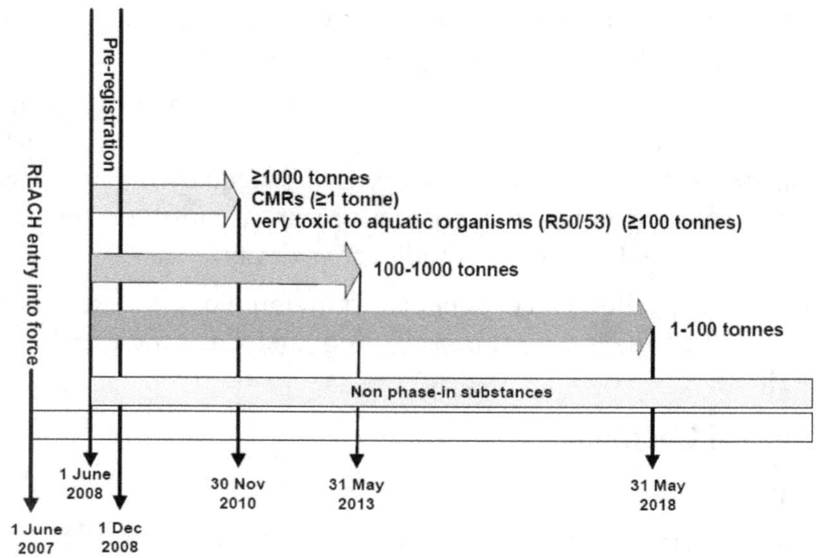

Figure 1. Scheme for the registration phase of the REACH regulation for existing industrial chemicals of different tonnage bands and hazardous potential, as well as newly developed chemicals ("Non phase-in") for the EU market. CMRs = chemicals that are proven carcinogenic, mutagenic or toxic to reproduction. R50/R53 labels indicate "Very toxic to aquatic organisms"/ "May cause long-term adverse effects in the aquatic environment". Source: http://www.cirs-reach.com/REACH/REACH_Registration_Deadlines.html (with permission).

In REACH, each chemical is registered only once. Accordingly, companies must work together to prepare one dossier that demonstrates to the European Chemical Agency (ECHA) how chemicals can be safely used, and they must communicate the risk management measures to the users. ECHA, or any Member State, authorizes the dossiers, and can start a "restriction procedure" when they are concerned that a certain substance poses an unacceptable risk to human health or the environment. If the risks cannot be managed, authorities can restrict the use of substances in different ways. In the long run, the most hazardous substances should be substituted with less dangerous ones.

So which chemicals have been registered in the past decade (2008-2018) in REACH? In principle, REACH applies to all chemical 'substances' in the EU zone. This includes metals, such as "iron" and "*chromium*", organic chemicals such as "methanol" and "fatty acids" and "ethyl-4-(8-chloro-5,6-dihydro-11H-benzo[5,6] cyclohepta[1,2-b]pyridin-11-ylidene)piperidine-1-carboxylate (see Box 1)", and (nano)particles like "zink oxide" and "silicon dioxide", and polymers. Discover for example the registration dossier link in Box 1.

Box 1. Examples from the REACH dossiers

The REACH registration data base can be searched via LINK. Accept the disclaimer, and you are ready to search for chemicals based on name, CAS number, substance data, or use and exposure data. Search for example for the name "ethyl 4-(8-chloro-5,6-dihydro-11H-benzo[5,6]cyclohepta [1,2-b] pyridin-11-ylidene)piperidine-1-carboxylate" and you find the link to the dossier of this substance with CAS 79794-75-5 as compiled by the registrant. This complex chemical name is better known as the antihistamine drug Loratadine, but this name does not show up in the dossier search!

Click on the name to get basic information on the compound. The hazard classification reads: "Warning! According to the classification provided by companies to ECHA in REACH registrations this substance is very toxic to aquatic life, is very toxic to aquatic life with long lasting effects, is suspected of causing cancer, causes serious eye irritation, is suspected of causing genetic defects, causes skin irritation, may cause an allergic skin reaction and may cause respiratory irritation." This compound is "PBT" labeled based on limited available data (classifying as a combination of Persistent / Bioaccumulative / Toxic). However, the section [About this substance] reads: "for industrial use resulting in the manufacture of another substance (use of intermediates)." As an intermediate in a restricted process, many parts of the dossier did not have to be completed for REACH. As a medicinal product Loratadine is strictly regulated elsewhere. Scroll down to the REACH link for the registration dossier (.../21649) to find out more the different entries for this chemical.

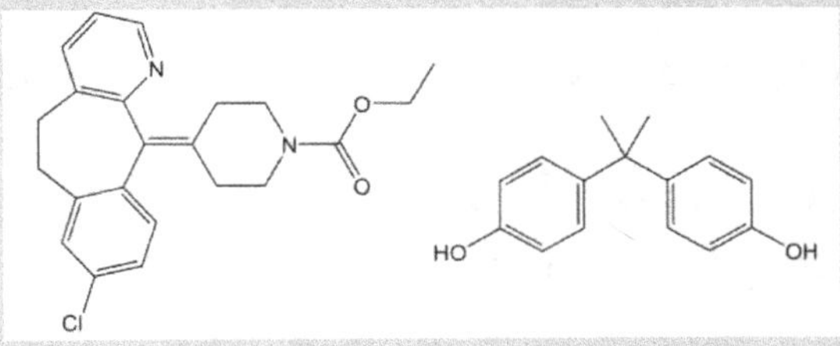

If we do a search for [" Bisphenol "], we get a long list of optional chemicals, for example Bisphenol A (CAS 80-05-7) but also for example Bisphenol S if you scroll down further (CAS 80-09-1). If we look at the dossier of the first Bisphenol A entry, with tonnage "100 000 - 1 000 000 tonnes per annum", you can find a long list of REACH information packages besides the dossier, as this chemical is hotly debated. The dossier for Bisphenol A was evaluated in 2013, and also this is available (look for the pdf in the Dossier evaluation status). In this compliance check, the registrant is requested to submit additional rat and mouse toxicity data, along with statements of reasons. There is for example also a link to the [Restriction list (annex XVII)], which leads to a pdf called 66.pdf, which states an adopted restriction for this chemical within the REACH framework and the previous legislation, Directive 76/769/EEC: "Shall not be placed on the market in thermal paper in a concentration equal to or greater than 0,02 % by weight after 2 January 2020".

Find your own chemical of interest to discover more on the transparancy of the chemical information on which risk assessment is based.

However, some groups of chemicals are (partly) exempt from REACH because they are covered by other legislation in the EU:

- Active substances used in plant protection products (Section 2.3.1) and biocidal products (Section 2.3.2) are considered as already having been registered and assessed by institutes separate from ECHA. Biocides such as disinfectants and pest control products are per definition hazardous chemicals, but they are also very useful in many ways. The very strict and elaborate biocide laws aim to verify that the potential risk of harm associated with the intended emission scenarios is in balance with expected benefits.
- Food and feedstuff additives (Section 2.3.9) have different legislation and authorisation laws to demonstrate (following a scientific evaluation) that the additive has no harmful effects on human and animal health or on the environment (developed since 1988, schematic graph, Regulation (EC) No 1331/2008)
- Medicinal products (Sections 2.3.3 and 2.3.4) have different legislation and authorisation laws to guarantee high standards of quality and safety of medicinal products, while promoting the good functioning of the internal market with measures that encourage innovation and competiveness (starting with Directive 65/65 in 1965, an overview since, a pdf of the 2001 EU legislation 2001/83/EC)
- "Waste" is not part of the REACH domain, but a product recovered from waste is not.

A detailed overview of European chemical safety guidelines related to chemicals with different application types is presented in Figure 2.

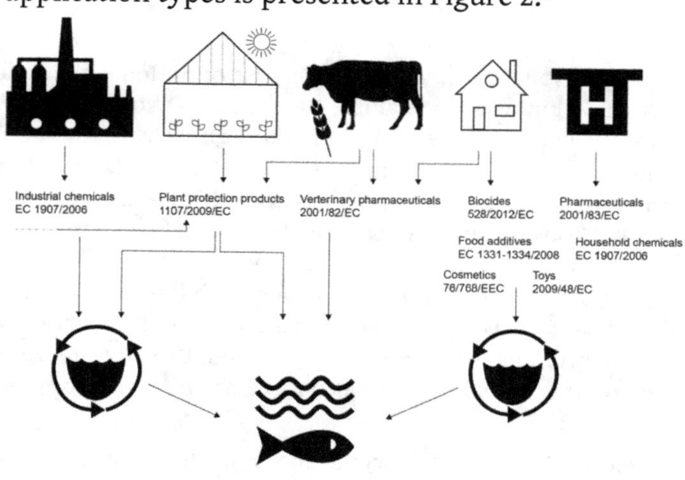

Figure 2. Scheme of societal sectors, their chemical uses, the dedicated policy frameworks for registration and authorization, and pathways to the aqueous environment direct or via industrial- or household effluent treatment plants (circle symbols). Redrawn from Van Wezel *et al.* (2017) by Evelin Karsten-Meessen.

Following pre-registration of the 145,297 chemicals most likely to require regulation, REACH came into force in 2008 in a stepwise process with different deadlines for different groups of chemicals. The first dossiers were to be completed by 2010 for the highest produced volume chemicals (>1000 tonnes/y) and the most hazardous chemicals (CMRs >1 tonne/y, and chemicals with known very high aquatic toxicity >100 tonnes/y). These groups potentially pose the greatest risk because of either their high emissions or their inherent toxicity. In 2013, registration dossiers for chemicals with a lower tonnage (100-1000 tonnes/y) were to be completed. By May 31 2018, all chemicals with a quantity of 1-100 tonnes/y chemicals on the EU market should have been registered. New chemicals will all be subject to the REACH procedures.

In 2018, 21.787 substances had been registered under REACH. A total of 14.262 companies were involved. In comparison, 15.500 substances were registered in 2016 (i.e., 6287 chemicals were added in the two following years). In 2018, 48% of all substance registrations had been done in Germany. For 24% of the registered substances a dossier was already available prior to REACH, 70% are "old chemicals" for which no registration had been done before REACH was initiated, and only 6% are newly developed substances that needed to be registered before manufacture or import could start.

There are multiple benefits of REACH regulation of industrial chemicals. Most data on chemicals entered in the registration process is publically available, creating transparency and improving customer awareness. If registered chemicals are classified as Substance of Very High Concern (SVHC) based on the chemical information in these dossiers and after agreement from research panels, alternatives that passed the same regulation can be suggested instead.

The necessity to add data on potential toxicity for so many chemicals has been combined with a strong focus on, and further development of, animal friendly testing methods. Read-across from related chemicals, weight of evidence approaches, and calculations based on chemical structures (QSAR) allow much experimental testing to be circumvented. In vitro studies are also used, but a 2017 REACH document (REACH alternatives to animal testing 2017, which followed up 2011 and 2014 reports) reports that 5.795 in vitro studies were used overall to determine endpoints for REACH, compared to 9,287 in vivo studies (ratio of 0.6).

Clearly, many new animal tests have been performed under REACH to complete the dossiers on industrial chemicals. Prenatal developmental and repeated dose toxicity testing as well as extended one generation reproductive toxicity studies remain difficult to circumvent without animal use. However, the safe use of industrial chemicals must be ensured and demonstrated.

References

Van Wezel, A.P., Ter Laak, T.L., Fischer, A., Bäuerlein, P.S., Munthee, J., Posthuma, L. (2017). Mitigation options for chemicals of emerging concern in surface waters; operationalising solutions-focused risk assessment . Environmental Science: Water Research 3, 403–414.

2.2.4. POPs

Introduction

Chemicals are generally produced because they have a useful purpose. These purposes can vary widely, such as to protect crops by killing harmful insects or fungi, to protect materials against catching fire, to act as a medicine, to enable a proper packing of food materials, etc. Unfortunately, the properties which make a chemical attractive to use, often have a downside when it comes to environmental behavior and/or human health. A number of synthetic chemicals have properties that make them persistent organic pollutants (POPs). POPs are xenobiotic (foreign to the biosphere) chemicals that are persistent, bioaccumulative and toxic ('PBT') in low doses. In addition, they are transported over long distances. Criteria for these properties, which are used to define a chemical as a POP, were set by the United Nations (UN) Stockholm Convention, which was adopted in 2001 and entered into force in 2004 (Fiedler et al., 2019). These criteria are summarized in Table 1 (http://chm.pops.int). The objective of the Stockholm Convention is defined in article 1: "Mindful of the precautionary approach, to protect human health and the environment from the harmful impacts of persistent organic pollutants". Initially, 12 chemicals (aldrin, chlordane, dieldrin, DDT, endrin, heptachlor, hexachlorobenzene (HCB), mirex, polychlorinated biphenyls (PCBs), polychlorinated dibenzo-p-dioxins (PCDDs), polychlorinated dibenzofurans (PCDFs) and toxaphene) were listed as POPs. Gradually the list was extended with new POPs, which appeared to fulfil the criteria. For some of the new chemicals exceptions were made for limited use, in case no suitable alternatives are available. For example, in the battle against malaria DDT can still be used to a limited extent for in-house spraying in Africa (Van den Berg, 2009). Until now all POPs are chemicals that contain carbon and halogen atoms. Some POPs, such as the PCDDs and PCDFs (together often short-named as dioxins) are not intentionally produced. They are formed and released unintentionally during thermal processes. PCDDs and PCDFs tended to be released by waste incinerators (Karasek and Dickson, 1987). The combination of elevated temperatures and the presence of chlorine from e.g. polyvinylchloride (PVC) led to the formation of the extremely toxic PCDDs and PCDFs. Stack emissions could contaminate entire areas around the incinerators with consequences for the quality of cow milk or local crop. Dioxins were first discovered after the Seveso (Italy) disaster (1976) when high quantities of dioxins were released after an explosion in a trichlorophenol factory (Mocarelli et al., 1991). Meanwhile, in many countries incinerators have been improved by changing the processes and installing appropriate filters.

Table 1. Stockholm Convention criteria for persistence, bioaccumulation, toxicity and long range transport of POPs.

Persistence	
i	Evidence that the half-life of the chemical in water is greater than two months, or that its half-life in soil is greater than six months, or that its half-life in sediment is greater than six months; or
ii	Evidence that the chemical is otherwise sufficiently persistent to justify its consideration within the scope of this Convention
Bioaccumulation	
i	Evidence that the bio-concentration factor or bio-accumulation factor in aquatic species for the chemical is greater than 5,000 or, in the absence of such data, that the log Kow is greater than 5
ii	Evidence that a chemical presents other reasons for concern, such as high bioaccumulation in other species, high toxicity or ecotoxicity; or
iii	Monitoring data in biota indicating that the bio-accumulation potential of the chemical is sufficient to justify its consideration within the scope of this Convention
Long range transport potential	
i	Measured levels of the chemical in locations distant from the sources of its release that are of potential concern
ii	Monitoring data showing that long-range environmental transport of the chemical, with the potential for transfer to a receiving environment, may have occurred via air, water or migratory species; or
iii	Environmental fate properties and/or model results that demonstrate that the chemical has a potential for long-range environmental transport through air, water or migratory species, with the potential for transfer to a receiving environment in locations distant from the sources of its release. For a chemical that migrates significantly through the air, its half-life in air should be greater than two days
Adverse effects	
i	Evidence of adverse effects to human health or to the environment that justifies consideration of the chemical within the scope of this Convention; or
ii	Toxicity or ecotoxicity data that indicate the potential for damage to human health or to the environment

Structures and Use

Whereas all initial POPs were chlorinated chemicals and mainly pesticides, POPs that were added at a later stage also included brominated and fluorinated compounds and chemicals with a more industrial application. Brominated diphenylethers (PBDEs) and hexabromocyclododecane (HBCD) belong to the group of brominated flame retardants. These chemicals are being produced in high quantities. Many national legislations require the use of flame retardants in many materials, such as electric and electronic systems (TV, cell phones, computers), furniture and building materials. Although the PBDEs and HBCD have been banned in most countries, other brominated flame retardants are still being produced in annually growing volumes.

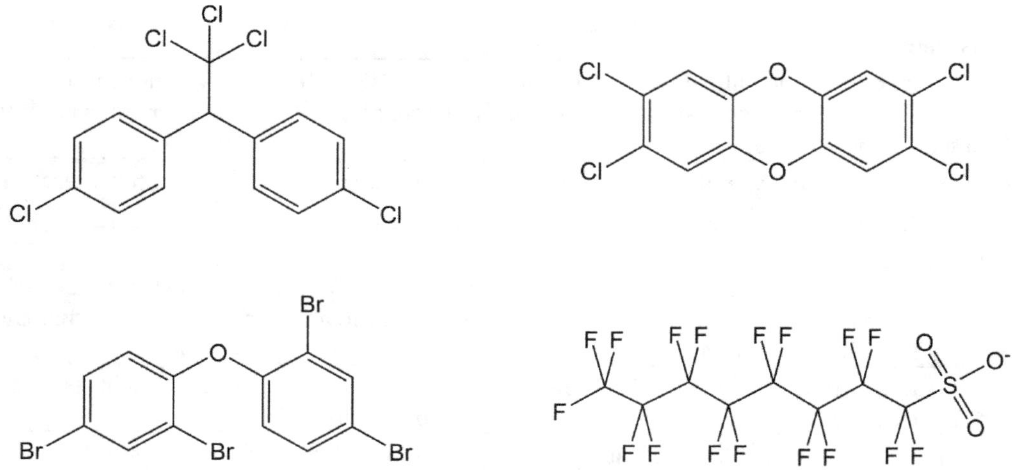

Figure 1. Structures of p,p'-DDT, 2,3,7,8-tetrachloro-p-dioxin, ,4,2',4'-tetrabromodiphenylether (a specific PBDE) and perfluoroctane sulfonic acid (PFOS).

Perfluorinated alkyl substances (PFASs) have many applications. Examples are Teflon production, use in fire-fighting foams, ski wax, as dirt and water repellant on outdoor clothes and carpets and many more. They are different from most other POPs because they are both lipophilic and hydrophilic due to a polar group present in most of the molecules. Examples of structures of a few POPs are given in Figure 1.

Persistence and Bioaccumulation

The carbon-halogen bond is so strong that any type of degradation is unlikely to occur or will only occur on the long term and to a minor extent. Due to the size of the halogen atom, the strength of the halogen-carbon bond decreases from C-F to C-Cl, C-Br and C-I. In addition, these halogenated chemicals are lipophilic and, therefore, easily migrate to lipids, such as in living organisms. Because fish is a primary target, POPs enter the food chain in this way and biomagnification can occur (De Boer et al., 1998). High levels of POPs are, consequently, found in marine mammals (seals, whales, polar bears) and also in humans (Meironyte, 1999). Women may transfer a part of their POP load again to their children, the highest quantities to their firstborns.

Long Range Transport

Chemicals that migrate significantly through the air with a half-life in air greater than two days qualify for the POP criterion of long range transport. Many chemicals are indeed transported by air, often in different stages. Chemicals are emitted from a stack or evaporate from the soil in relatively warm areas and travel in the atmosphere toward cooler areas, condensing out again when the temperature drops. This process, repeated in 'hops', can carry them thousands of kilometers within days. This is called the 'grasshopper effect' (Gouin et al., 2004). It results in colder climate zones, in particular countries around the North Pole, receiving relatively high amounts of POPs.

Adverse Environmental and Health Effects

There is very little doubt on the toxicity of POPs. Of course, the dose is always determining if a compound is causing an effect in the environment or in humans. POPs, however, are very toxic at very low doses. The Seveso disaster showed the high toxicity of dioxins for humans. Polybrominated biphenyls (PBBs) caused a high mortality in cattle when they were inadvertently fed with these chemicals (Fries and Kimbrough, 2008). Evidence of toxicity is often coming from laboratory studies with animals (*in vivo*) and more recently from *in vitro* studies. These studies are in particular important for the assessment of chronic toxicity. Many POPs are carcinogenic or act as endocrine disruptor.

Analysis

The analysis of POPs in environmental or human matrices is relatively complicated and costly. The compounds need to be isolated from the matrix by extraction. Subsequently, the extracts need to be cleaned from interfering compounds such as fat from organisms or sulphur in case of sediment or soil samples. Finally, due to the required sensitivity and selectivity, expensive instrumentation such as gas or liquid chromatography combined with mass spectrometry is needed for their analysis (Muir and Sverko, 2006). UN Environment is investing in large capacity building programs to train laboratories in developing countries in this type of analysis. According to the Stockholm Convention, countries shall manage stockpiles and wastes containing POPs in a manner protective of human health and the environment. POPs in wastes are not allowed to be reused or recycled. A global monitoring program has been installed to assess the effectiveness of the Convention.

Future

Much has to be done to achieve the original goals of eliminating the production and use of POPs and gradually reduce their spreading into the environment. A global treaty such as the Stockholm Convention with 182 countries involved is in a continuous challenge with procedures and political realities in countries, which hamper the achievement of perceived simple goals such as to eliminate the use of PCBs in 2025. The goals are, however, extremely important, as POPs are a global threat for current and future generations.

References

De Boer, J., Wester, P.G., Klamer, J.C., Lewis, W.E., Boon, J.P. (1998). Brominated flame retardants in sperm whales and other marine mammals - a new threat to ocean life? Nature 394, 28-29.

Fiedler, H., Kallenborn, R., de Boer, J., Sydnes, L.K. (2019). United Nations Environment Programme (UNEP): The Stockholm Convention - A Tool for the global regulation of persistent organic pollutants (POPs). Chem. Intern. 41, 4-11.

Fries, G.F., Kimbrough, R.D. (2008). The PBB episode in Michigan: An overall appraisal. CRC Critical Rev. Toxicol. 16, 105-156.

Gouin, T., Mackay, D., Jones, K.C., Harner, T., Meijer, S.N. (2004). Evidence for the "grasshopper" effect and fractionation during long-range atmospheric transport of organic contaminants. Environ. Pollut. 128, 139-148.

Karasek, F.W., Dickson, L.C. (1987). Model studies of polychlorinated dibenzo-p-dioxin formation during municipal refuse incineration. Science 237, 754-756.

Meironyte, D., Noren, K., Bergman, Å. (1999). Analysis of polybrominated diphenyl ethers in Swedish human milk. A time-related trend study, 1972-1997. J. Toxicol. Environ. Health Part A 58, 329-341.

Mocarelli,P., Needham,L.L.,Marocchi,A.,PattersonJr.,D.G.,Brambilla,P.,Gerthoux, P.M. (1991). Serum concentrations of 2,3,7,8-tetrachlorodibenzo-p-dioxin and test results from selected residents of Seveso, Italy. J. Toxicol. Environ. Health 32, 357-366.

Muir, D.C.G., Sverko, E. (2006). Analytical methods for PCBs and organochlorine pesticides in environmental monitoring and surveillance: a critical appraisal. Anal. Bioanal. Chem. 386, 769-789.

Van den Berg H. (2009). Global status of DDT and its alternatives for use in vector control to prevent disease. Environ. Health Perspect. 117:1656–63.

2.2.5. Persistent Mobile Organic Chemicals (PMOCs)

Introduction

Ecosystems and humans are protected against exposure to hazardous substances in several ways. These include treating our wastewater so that substances are prevented from entering receiving surface waters, and purification of source waters intended for drinking water production. Currently, a majority of the drinking water produced in Europe is either not treated or treated by conventional technologies. The latter remove substances by degradation (physical, microbiological) or by sorption. However, chemicals that are difficult to break down and that can pass through soil layers, water catchments and riverbanks and cross natural and technological barriers may eventually reach the tap water. Typically, these chemicals are persistent and mobile.

Polarity

When the electrons in a molecule are unevenly divided over its surface, this results in an asymmetric distribution of charge, with positive and negative regions. Such molecules have electric dipoles (see Figure 1) and are polar, in contrast to molecules where the charge is evenly distributed thus resulting in the molecule being neutral or apolar. The ultimate form of polarity is when a permanent charge is present in a compound. Such chemicals are called ionogenic. We distinguish between cations (having a permanent positive charge, e.g. protonated bases, and quaternary amines) and anions (negatively charged ions, e.g. dissociated acids, and organosulfates).Ionic charges in molecules can be pH dependent (e.g. acids and bases). Most, and in particular small, polar and ionic chemicals are water soluble, in other words they have a strong affinity to water (often referred to as hydrophilic). Because water is one of the most polar liquids possible (a strong negative charge on the oxygen and two strong positive charges on each hydrogen), this means that for very polar organic molecules solvation by water is more favorable energetically then sorption to solid particles.

Chemicals that are nonpolar are inherently poorly water soluble and therefore tend to escape from the water compartment, resulting in their evaporation, or sorption to sediments and soils, or uptake and accumulation in organisms. It is therefore relatively easy to remove them from water during water treatment. In contrast, mobile organic chemicals, especially those that do not breakdown easily, pose a more serious threat for (drinking) water quality because they are much more difficult to remove. It should be noted that mobility and polarity can be thought of as a gradient, rather than a distinct category, with water being the most polar molecule, a large aliphatic wax being the most non-polar molecule, and all other organic molecules falling somewhere in the spectrum between.

In a recent study contaminants were analysed in Dutch water samples covering the journey from WWTP effluent to surface water to groundwater and then to drinking water. While the concentration level of total organic contaminants decreased by about 2 orders of magnitude from the WWTP effluents to the groundwater used for drinking water production, the hydrophilic contaminants (using chromatographic retention time as an indicator for hydrophilicity) in the WWTP effluents remained in the water throughout its passage to groundwater and into the drinking water (see Figure 2).

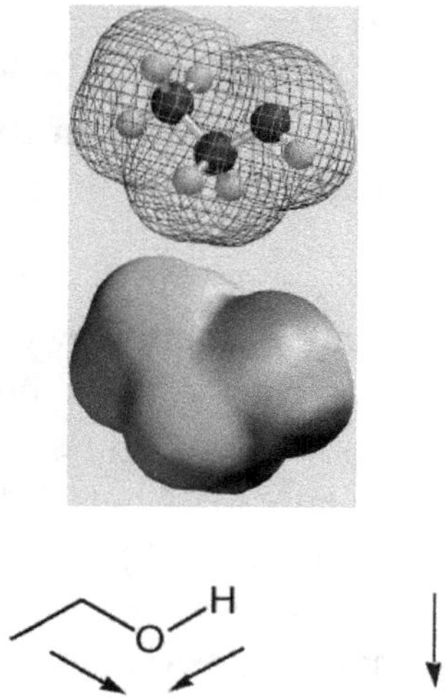

Figure 1. A model of ethanol showing the volume that is occupied by its electrons (A) and the direction of the dipole (B). The distribution of the electrons in an ethanol molecule is skewed relative to the protons to give a region having a partial negative charge, which is shown in red, and a corresponding region having a partial positive charge, which is shown in blue. (source: https://www.chem.uwec.edu/Chem150_Resources/content/elaborations/unitx/unit1-e-polarity.html and http://www.utdallas.edu/~biewerm/10-alcohols.pdf)

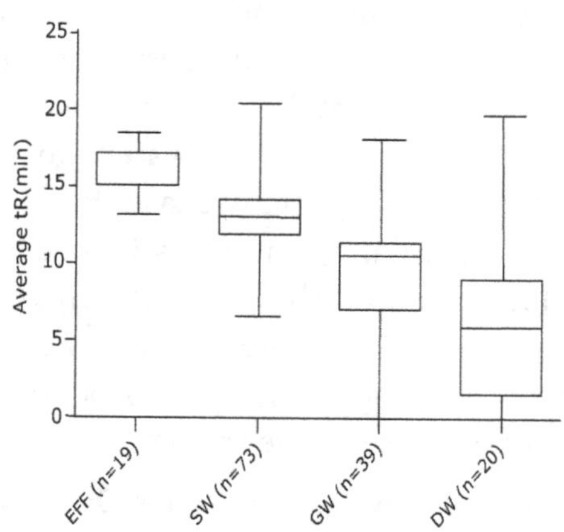

Figure 2. Average chromatographic retention time (tR) as a measure of average hydrophilicity of contaminants present in different water types. EFF, effluent; SW, surface water; GW, groundwater; DW, drinking water. Redrawn from Sjerps et al. (2016) by Wilma IJzerman.

Mobility and Persistence

The mobility of chemicals in aquatic ecosystems is determined by their distribution between water and solid particles. The more the substance has an affinity for the solid phase the less mobile it will be. The distribution coefficient is known as $K_{D,}$ which expresses the ratio between the concentrations in the solid phase (soil, sediment, suspended particles), C_s, and the dissolved phase at equilibrium, C_W, i.e. $K_D = C_s/C_W$. For neutral non-polar chemicals the distribution is almost entirely determined by the amount of organic carbon in the solid phase, f_{OC}, and hence their distribution is usually expressed by K_{OC}, the organic carbon-normalized K_D (i.e. $K_{OC} = K_D/ f_{OC}$). Unfortunately, there are relatively few reliable K_D or K_{OC} data available, in particular for polar chemicals. Instead, K_{OW} is often used as a proxy for K_{OC}. The n-octanol/water partition coefficient: K_{OW}, is the equilibrium distribution coefficient of a chemical between n-octanol and water, $K_{OW} = C_{octanol}/C_{water}$. It's logarithmic value is often used as a proxy to express the polarity of a compound: a high log K_{OW} means that the compound favors being in the octanol phase rather than in water, which is typically the case for a nonpolar compound. For ionizable chemicals we need to account for the pH dependency of K_{OW}: at low pH an organic acid will become protonated (this in turn depends on its pK_a value) and thus less polar. D_{OW} is the pH-dependent K_{OW}. It can be assumed that ions, whether cationic or anionic, will no longer dissolve into octanol but rather be retained in the water, because ions have much higher affinity for water than for octanol.

Accounting for this, for organic acids, the pH dependency of the D_{OW} can be expressed as:

$$D_{OW} = \frac{K_{OW}}{1 + 10^{\,pH - pK_a}}$$

Therefore, as pH increases above the pKa, the smaller the D_{OW} will get in the case of organic acids. In the case of basis, the opposite is true; the more the pH dips below the pKa of an organic base, the more cations form and the lower the Dow becomes.

However, one has to keep in mind that the assumption that the (log) K_{OW} or D_{OW} value inversely correlates with a compound's aquatic mobility is, certainly, very simplistic. The behavior of an ionic solute will obviously also be determined by interactions i) with sites other than organic carbon, e.g. ionizable or ionic sites on soil and sediment particles, and ii) with other ions in solution.

The persistence of a compound is assessed in experimental tests by monitoring the rate of disappearance of the compound from the most relevant compartment. This is often done using standardized test protocols. In the European REACH legislation of chemicals, criteria have been established to qualify chemicals as persistent (P) and "very persistent"(vP) based on the outcomes of such tests. Table 1 presents the P and vP criteria used. Unfortunately, good-quality experimental data on half-lives are rare and obtaining such data requires both time-consuming and expensive testing.

Currently there is no certified definition of a compound's mobility (M). Several possible compound properties have been proposed to characterize mobility, including a compound's aqueous solubility or its K_{oc} value. If (experimental) K_{oc} values are not available, D_{ow} values can be used as a proxy.

Table 1. P and vP criteria identical to Annex XIII to the REACH regulation (source: ECHA chapter R.11. Version 3.0, June 2017)

	Persistent (P) in any of the following situations	Very persistent (vP) in any of the following situations
Freshwater	Half-life > 40 days	Half-life > 60 days
Marine water	Half-life > 60 days	Half-life > 60 days
Freshwater sediment	Half-life > 120 days	Half-life > 180 days
Marine sediment	Half-life > 180 days	Half-life > 180 days
Soil	Half-life > 120 days	Half-life > 180 days

Table 2. Proposed cut off values of compound properties proposed by the German Environmental Agency (UBA) to define substance mobility*

	Mobile (M) if compound is P or vP and any of the following situations	very Mobile (vM) if compound is P or vP and any of the following situations
Lowest experimental log KOC (at pH 4-9)	≤4.0	≤3.0
Log DOW (at pH 4-9)	≤4.0 if no experimental Log KOC data available	≤3.0 if no experimental Log KOC data available

* note that the proposed criteria may change by the time of publication

Regulation and Gaps in Knowledge

The majority of chemicals for which international guidelines exist or that are identified as priority pollutants in existing regulations (e.g. EU Water Framework Directive and REACH), are nonpolar with log DOW values mostly above two (see Figure 3b). The German Ministry of Environment (UBA) has recently proposed to develop regulation for chemicals with P, M and toxic (T) properties (PMT substances) analogous to the existing PBT criteria used for regulation of chemicals in the EU. UBA proposed to use cut-off values of the K_{OC} or D_{OW} (if K_{OC} data are not available) to define Mobile (M) or very mobile (vM) in conjunction with persistence criteria (see Table 2). Note that the KOC and DOW values have to be obtained from testing at an environmentally relevant pH range (pH 4-9).

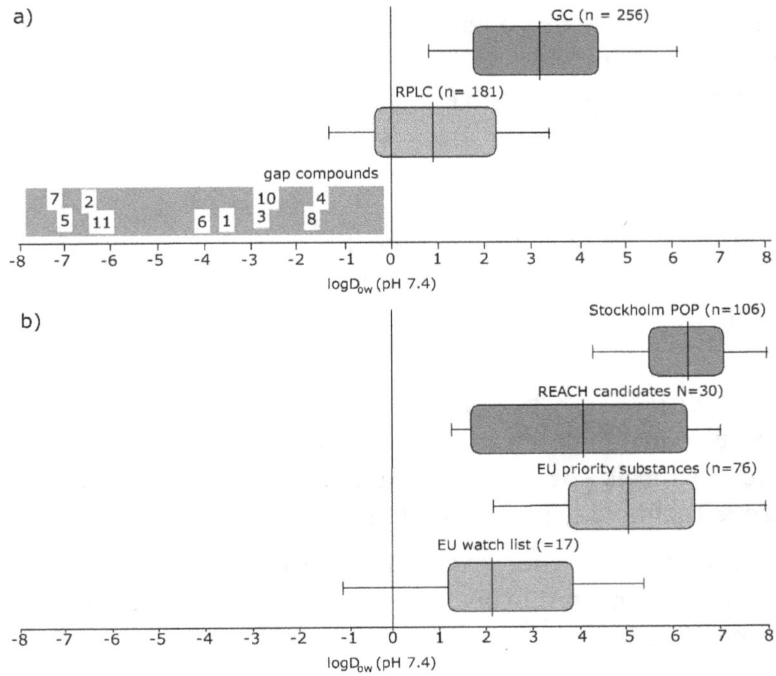

Figure 3. Box and whisker plots of calculated logDOW values at pH 7.4 of: (a) contaminants in water analyzed by either GC-MS or LC-MS and of examples of mobile chemicals; (b) contaminants regulated by the Stockholm Convention; candidates of Substances of Very High Concern (SVHCs) according to REACH, Article 57 d-f; the list of priority substances according to the Water Framework Directive (WFD); and the so-called Watch List of the WFD. The whiskers point to the 10th and 90th percentile. Numbers in (a) refer to 1: Aminomethylphosphonic acid (AMPA), 2: Paraquat, 3: Cyanuric acid, 4: N,N-dimethylsulfamide (DMS), 5: Diquat, 6: 5-Fluorouracil, 7: Glyphosate, 8: Melamine, 9: Metformin, 10: Perfluoroacetic acid, 11: EDTA. Redrawn from Reemtsma et al. (2016) by Wilma IJzerman.

When we consider current analytical techniques used for monitoring contaminants in the environment, it can be readily seen that the scope of techniques most often used (gas chromatography, GC, and reversed-phase liquid chromatography, RPLC) do not overlap with what is required for chemicals having log DOW values typical of the most mobile chemicals, having a log D_{ow} below zero (see Figure 3a). Consequently, there is limited information available on the occurrence and fate of these mobile chemicals in the environment. Nevertheless, some examples of persistent and mobile chemicals have been identified. These include highly polar pesticides and their transformation products, for instance glyphosate and aminomethylphosphonic acid (AMPA), short-chain perfluorinated carboxylates and sulfonates, quaternary ammonium chemicals such as diquat and paraquat and complexing agents such as EDTA. There are, however, likely to be many more chemicals that could be classified as PMOC and we can therefore conclude that there is a gap in the knowledge and regulation of persistent mobile organic chemicals.

References

Arp, H.P.H., Brown, T.N., Berger, U., Hale, S.E. (2017). Ranking REACH registered neutral, ionizable and ionic organic chemicals based on their aquatic persistency and mobility. Environmental Science: Processes Impacts 19, 939-955.

Reemtsma, T., Berger, U., Arp, H. P. H., Gallard, H., Knepper, T. P., Neumann, M., Benito Quintana, J., de Voogt, P. (2016). Mind the gap: persistent and mobile organic chemicals - water contaminants that slip through. Environmental Science & Technology 50, 10308-10315.

Sjerps, R.M.A., Vughs, D., van Leerdam, J.A., ter Laak, T.L., van Wezel, A.P. (2016). Data-driven prioritization of chemicals for various water types using suspect screening LC-HRMS. Water Research 93, 254-264.

2.2.6. Ionogenic Organic Chemicals

Introduction

Ionogenic organic chemicals (IOCs) are widely used in industry and daily life, but also abundantly present as chemicals of emerging concern. For environmental risk assessment purposes, IOCs may be defined as organic acids, bases, and zwitterionic chemicals that under common environmental pH conditions exist for a large part as charged (ionic) species, with only modest fraction of neutral species. The environmentally relevant pH range could be argued to lie between 4 (acidic creeks, even lower in polluted streams from volcanic regions or mine drainage systems) to 9 (sewage treatment effluents). The environmental behaviour of IOC pollutants of concern are different from neutral chemicals of concern, because the aqueous pH controls the neutral fraction of dissolved IOCs and the ionic form is highly soluble and interacts partly via electrostatic interactions with environmental susbtrates. Note that the major fraction of an IOC can also be neutral in a certain environmental system, and in that case it is often the neutral form that dominates the chemical's behavior.

IOCs are common in many different types of pollutant classes. A random subset analysis of all EU (pre-)registered industrial chemicals indicated that large fractions of the total list of >100,000 chemicals are IOCs (51% neutral; 27% acids;

14% bases; 8% zwitterions/amphoterics). In another source, It has been estimated that >60% of all prescription drugs (Section 2.3.3) are IOCs (Manallack, 2007), and even higher fractions for illicit drugs (Section 2.3.4) (Figure 1). Well known examples are basic beta-blockers (e.g. propranolol), basic antidepressants (e.g. fluoxetine and sertraline), acidic non-steroidal anti-inflammatory drugs (NSAID such as diclofenac), and basic opioids (e.g. morphine, cocaine, heroin) and basic designer drugs (e.g. MDMA). The majority of surfactants and polyfluorinated chemicals (e.g. PFOS and GenX) are IOCs (Section 2.3.8), as well as wide variety of important pesticides (e.g. zwitterionic glyphosate) (Section 2.3.1) and (natural) toxins (Section 2.1) (e.g. peptide based multi-ionic cyanobacterial toxins).

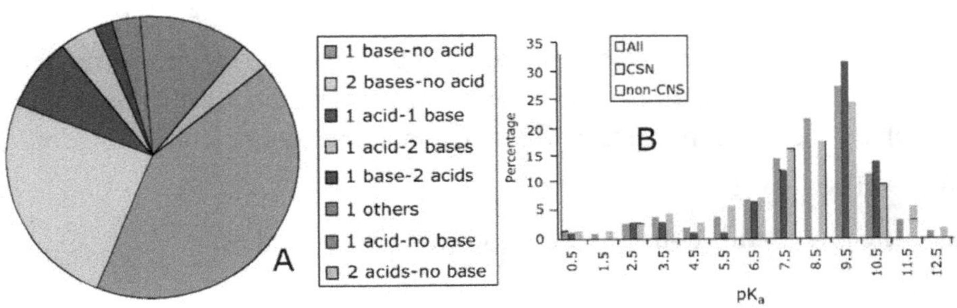

Figure 1. A survey on the 1999 World Drug Index (WDI) database revealed that >63% of the 51,596 chemicals had acidic or basic functionalities. Shown in the circle diagram A is the distribution of different ionizable groups in this fraction of 63% drugs, and in bar diagram B the distribution in pKa values for basic drugs (CNS = drugs acting on central nervous system). Adapted from Manallack (2007) by Wilma IJzerman.

Environmental Behavior of IOCs

The release into the environment is specific for each of these types of IOCs with a different use, but in many cases happens via sewage treatment systems. If sorption to sewage sludge is very strong, application of sludge onto terrestrial (agricultural) systems is a key entry in many countries. However, many IOCs are rather hydrophilic and will mainly be present in wastewater effluent released into aquatic systems. As they are hydrophilic, they are considered rather mobile which allows for rapid transport through e.g. groundwater plumes, soil aquifers, and (drinking water) filtration steps. The distinction between (mostly) neutral chemicals and IOCs is important because the ionic molecular form generally behaves very differently from the corresponding neutral molecular form. For example, in many aspects ionic molecules are non-volatile compared to the corresponding neutral molecules, while neutral molecules are more hydrophobic than the corresponding ionic molecules. As a result of their lower "hydrophobicity", ionic molecules often bind with lower affinity to soils and are therefore more mobile in the environment. The ionic forms bioaccumulate to a lower extent and can therefore be less toxic than the corresponding neutral form (though not necessarily). However, there are various important exceptions for these rules. For example, clay minerals sorb cationic IOCs fairly strongly via ion exchange mechanisms. Certain proteins (e.g. the blood serum protein albumin) tightly bind anionic chemicals because of cationic subdomains on specific (enzymatic) pockets, which allows for effective transport throughout our systems and over cell membranes.

Calculating and Predicting the Dissociation Constant (pKa)

The critical chemical parameter describing the ability to ionize is the acid dissociation constant (pK_a). The pK_a defines at which pH 50% of the IOC is in either the neutral or ionic form by releasing an H+ from the neutral molecule acids (AH to anion A-), or accepting an H+ onto the neutral molecule base (B to cation BH+). The equilibrium between neutral acid and dissociated form can thus be defined as:

$$[HA] \leftrightarrow [H+] + [A-] \qquad \text{(eq.1)}$$

where the chemical's equilibrium speciation is defined as:

$$K_a = \frac{[H^+][A^-]}{[HA]} \qquad \text{(eq.2)}$$

which gives the pK_a as:

$$pKa=-\log(Ka) \qquad \text{(eq.3)}$$

and as a function of pH, the ratio of the acid and anion is defined as:

$$pH = pK_a + log \frac{[A^-]}{[HA]} \text{ for acids and for bases} \qquad \text{(eq.4)}$$

Although the term pK_b is also used to denote the base association constant, it is conventional we consider $[BH^+]$ as acid and use 'pK_a' and other relationships for bases as well. The fraction of neutral species (f_N) for simple IOCs (one acidic or basic site) can be readily calculated with a derivatization of the Henderson-Hasselbalch equation:

$$f_N = \frac{1}{1+10^{\alpha(-pH+pK_a)}} \qquad \text{(eq.5)}$$

in which α = 1 for bases, and -1 for acids.

In which α = 1 for bases, and -1 for acids. Using equation 5, Figure 2 presents a typical speciation profile for an acid (shown with pKa 5, so perhaps a carboxylic acid) and a base (shown with pKa 9, so perhaps a beta-blocker drug). Following the curve of equation 5, it is interesting to see some simple rules: if the pH is 1 unit lower than the pKa, the deprotonated species fraction is present at 10%. If the pH is

2 units lower than the pKa, the deminished species fraction is present at 1%, 3 units lower would give 0.1%, etc. From this, it is easy to make a good estimate for the protonation of a strong basic drug like MDMA (pKa reported 9.9-10.4) in blood (pH 7.4): with a maximum of 3 units lower pH, up to 99.9% of the MDMA will be in the protonated form, and only 0.1% neutral. For toxicological modeling studies, e.g. in terms of permeation through the blood-brain barrier membrane, this is highly relevant.

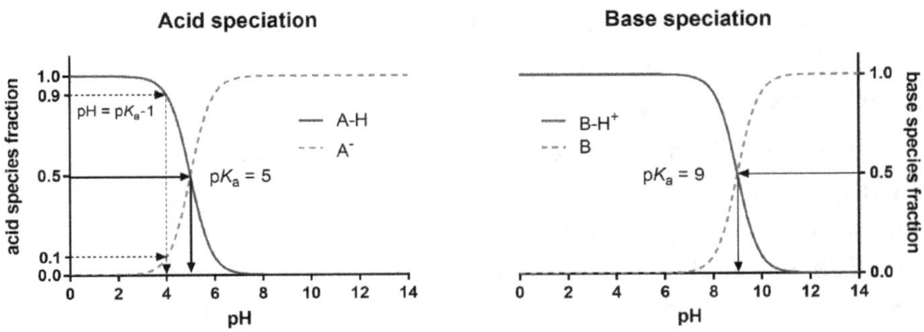

Figure 2. pH dependent speciation of an acid (with a pKa of 5) and a base (with a pKa of 9). As shown by the arrows, if pH = pK_a, the IOCs exist 50% in the neutral form, and 50% in the ionic form. For the acid at pH = pKa -1 (pH 4), 90% is in the neutral form (AH), and 10% is in the negatively charged form (A-). Drawn by Steven Droge.

Boxes 1-3. Extended learning: calculating the dissociation constant for multiprotic chemicals: see end of this module

Acidic IOCs:

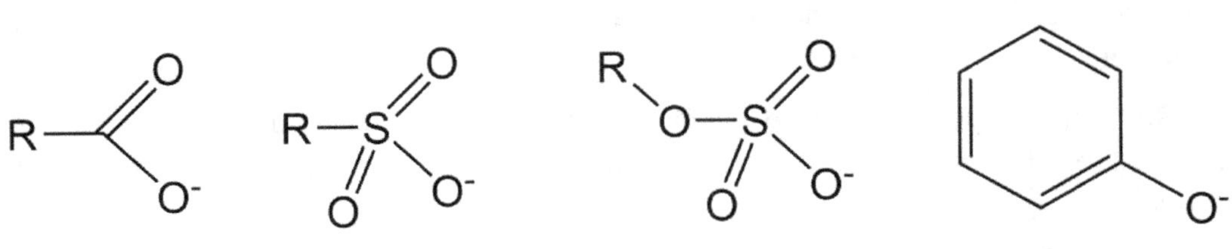

Figure 3. Different types of anionic moieties: carboxylate (anionic form of carboxylic acid), sulfonate (anionic form of sulfonic acid), sulfate (anionic form of sulfuric acid), phenolate (anionic form of phenol). (Source: Steven Droge)

For example, the painkiller (or non-steroidal anti-inflammatory drug, or NSAID) diclofenac is a carboxylic acid with a pKa of 4.1. This means that at pH 4.1, 50% of the dissolved diclofenac is in the dissociated (anionic) form (so, $(1 - f_N)$ from equation 5). At pH 5.1 (1 unit higher than the pKa) this is roughly 90% (90.91% to be more precise, but simply remembering 90% helps), at pH 6.1 (2 units higher than the pKa) this is 99%. This stepwise increase in 50-90-99% with each pH unit works for all acids, and for bases the other way round. Test for yourself that at physiological pH of 7.4 (e.g. in blood) diclofenac is calculated to be 99.95% anionic.

Many carboxylic acids have a pK_a in the range of 4-5, but the neighboring molecular groups can affect the pK_a. Particularly electronegative atoms such as chlorine, fluorine, or oxygen may lower the pKa, as they reduce the forces holding the dissociating proton to the oxygen atom. For example, trichloroacetic acid (CCl_3-COOH) has a pK_a of 0.77, while acetic acid (CH_3-COOH) has a pKa of 4.75. For the same reason, perfluorinated carboxylic acids have a strongly reduced acidic pKa compared to the analogous non-perfluorinated carboxylic acids.

Sulfate acids (see figure 3) are very strong acids, with a pKa <0. These acids almost always occur in their pure form as a salt, for example the common soap ingredient sodium dodecylsulfate ("SDS" or "SLS", $Na.C_{12}H_{25-}SO_4$). Other common detergents are sulfonates, such as linear alkylbenzenesulfonate ("LAS", C_{10-14}-(benzyl-SO_3)), where the SO_3 anionic moiety is attached to a benzene ring, which can be positioned to different carbon atoms of a long alkyl chain.

Even at the lower environmental pH range of about 4 these soap chemicals are fully in the anionic form. Such very strong acids, but also many weaker acids, are thus often sold in pure form as salts with sodium, potassium, or ammonium, which causes them to have different names and CAS numbers (e.g. $Na.C_{12}H_{25}$-SO_4 or $K.C_{12}H_{25}$-SO_4) than the neutral form.

Many phenols have a pK_a > 8, and are therefore mostly neutral under environmental pH. Electron-withdrawing groups on the aromatic ring of the phenol group, such as Cl, Br and I, can lower the dissociation constant. For example, the dinitrophenol-based pesticide dinoseb has a pK_a of 4.6, and is thus mostly anionic in the aquatic environment.

Note that a hydroxyl group (-OH) not connected to an aromatic ring, such as -OH of alcohols, can in most cases for risk assessment be considered permanently neutral.

To help interpret the differences in pKa between molecules, it sometimes helps to remember that in more acidic solutions, there are simply higher H^+ concentrations, in a logarithmic manner on the pH scale. At pH 3, the concentration H^+ protons in solution is 1 mM, while at pH 9 the H+ concentration is 1 nM (6 pH units equals 106 times lower concentrations). The affinity ("K_a") of H^+ to associate with a negatively charged molecular group, is so low for strong acids that even at very high dissolved H+ concentrations (low pH) only very few AH bonds (neutral acid fraction) are actually formed.

In other words, for chemicals with a low Ka, even at low pH the neutral fraction is still low. For weak acids such as phenols, already at a very low dissolved H+ concentrations (high pH) many AH bonds (neutral acid fraction) are formed.

So it can be reasoned that the affinity of common acidic groups to hold on to a proton is in the order:

$SO_4 < SO_3 < CO_2 <$ amide (C(=O)NH) < phenol < hydroxyl

Basic IOCs:

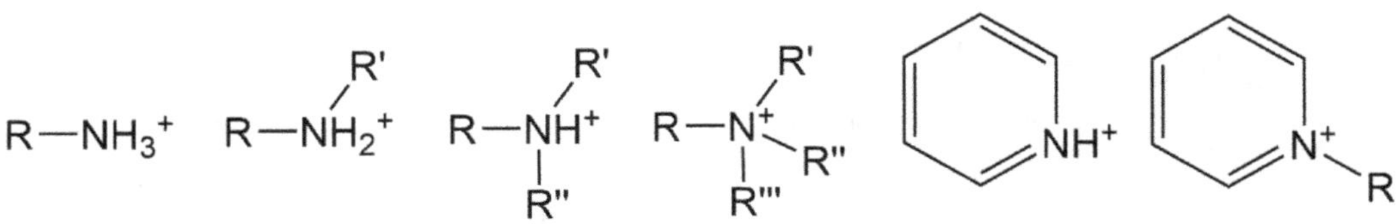

Figure 4. Different types of cationic moieties: primary amine, secondary amine, tertiary amine, quaternary ammonium (permanently charged), pyridinium (cationic form of pyridine), alkylpyridinium (permanently charged). (Source: Steven Droge).

For bases, it is mostly a nitrogen atom that can accept a proton to form an organic cation, because of the lone electron pair in nitrogen. Neutral nitrogen atoms have opportunities for 3 bonds. A primary amine group has the nitrogen atom bonding to only 1 carbon atom (represented here as part of a molecular fragment R), and two additional bonds with hydrogen atoms.

The remaining electron pair readily accepts another proton to form a cationic molecule [R-NH^{3+}]. Neutral secondary amines have one bond to hydrogen and two bonds to carbon atoms and can accept a proton to form [R-NH^{2+}-R'], whereas neutral tertiary amines have no bonds to hydrogen but only to carbon atoms and can form [R-NH^{+}-(R')(R'')]. Of course, each R group may be the same (e.g. a methyl unit).

Many basic chemicals have complex functionalities that can influence the pKa of the nitrogen moiety. However, as shown in the examples of figure 5, as long as there are at least two carbon atoms between the amine and the polar molecular fragment (for example OH, but much stronger for =O), the pKa of the basic nitrogen groups in all three types of bases (primary, secondary and tertiary amines) is high, often above 9 (dissolved H^+ concentration <10-9).

So even at very low H^+ concentrations, dissolved protons like to be associated to such amine groups. As a result, amines such as most beta-blockers and amphetamine based drugs are predominantly positively charged molecules (organic cationic amines) in the common environmental pH range of 4-9, as well as in the pH of most biotic tissues that are useful for toxicological assessments. As soon as a polar group with oxygen (e.g. ketones or hydroxyl groups) is connected to the second carbon away from the nitrogen (e.g. R-CH(OH)-CH_2-NH_2) the pK_a is considerably lowered.

Also nitrogen atoms as part of an aromatic ring, or connected to an aromatic ring, have much lower pK_a's: protons have rather low affinity to bind to these N-atoms and only start doing so if the proton concentration becomes relatively high (solution becomes more acidic).

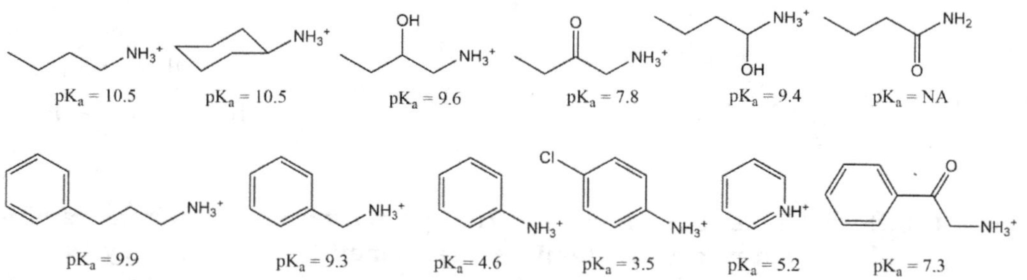

Figure 5. Different proton dissociation constants for amine groups: pKa is influenced by other functional structures. (Source: Steven Droge)

Relevance of Accounting for Electrostatic Interactions

Most classical pollutants, such as DDT, PCBs and dioxins, are neutral hydrophobic chemicals. On the other hand, most metals are almost always cationic species (e.g. Cd^{2+}). Consequently, their environmental distribution and biological exposure are influenced by quite distinct processes. Obviously, predominantly charged IOCs behave somewhat in between these two extremes. The charged positive or negative groups cause a strong effect of electrostatic interactions between the IOC and environmental substrates (sorption or ligand/receptor binding). While also metals speciate into different forms, pH difference between environmental compartments can strongly influence the IOCs chemical fate and effect if ionizable group is relatively weak. An important difference to metals is that the nonionic molecular part still influences the IOC's hydrophobicity even in charged state for several processes.

As will be discussed in other chapters regarding chemical processes (see Chapter 3), it needs to be taken into account for IOCs that many environmental substrates (DOC, soil organic matter, clay minerals) are mostly negatively charged in the common range of environmental pH, but also that proteins involved in biotic uptake-distribution-effects are rich in ionogenic peptides that are part of binding pockets and reactive centers.

References

Manallack, D.T. (2007). The pKa distribution of drugs: application to drug discovery. Perspectives in Medical Chemistry 1, 25-38.

Box 1. Extended learning: calculating the dissociation constant for multiprotic chemicals:

- Several common inorganic acids are multiprotic: they have multiple protons that can dissociate.
- Multiples species can occur at a certain pH, such as for phosphoric acid (H_3PO_4, $H_2PO_4^-$, HPO_4^{2-} , HPO_4^{2-}), and carbonic acid (H_2CO_3, HCO_3^-, CO_3^{2-}). It is important to realize that there are actually two micro-species of HCO_3^-, because two hydroxylgroups can dissociate: HO-C(=O)-OH

A polyprotc acid H_nA can undergo *n* dissociations to form *n*+1 species. Each dissociation has a pK_a.

But how to calculate the fraction of each species of multiprotic chemicals?

The charge of a polyprotic acid can be described as $H_{n-j}A^{j-}$. A useful variable, *v*, can be defined for each general polyprotic acid:

$$v = \frac{(H_nA]}{[H^+]^n} \; so \; \frac{([H_nA])}{v} = [H^+]^n$$

for each dissociation reaction:

$H_{n-j+1}A^{-j+1} \leftrightarrow H^+ + H_{n-j}A^{j-}$

the dissociation constant K_j is:

$$K_j = \frac{[H^+][H_{n-j}A^{J-}]}{[H_{n-j+1}A^{-j+1}]}, and\; pK_j = -log(K_j)$$

The degree of dissociation of the acid (η) is equal to the ratio of the total charge (TC) to the total mol acid (TM). For a diprotic acid, a plot of η as a function of pH provides the dissociation curve:

$$\eta = \frac{TC}{v}\frac{v}{TM} = \frac{TC}{TM} = \frac{1K_1[H^+]+2K_1K_2}{[H^+]^2+K_1[H^+]+K_1K_2}$$

which can be a fitted curve to experimental data.

The degree of protolysis for the jth species, α_j, can be calculated from the ratio of $[H_{n-j}A^{j-}]/TM$

$$\alpha_0 = \frac{[H_2A]}{v}\frac{v}{TM} = \frac{[H^+]^2}{[H^+]^2+K_1[H^+]+K_1K_2}$$

$$\alpha_1 = \frac{[HA^-]}{v}\frac{v}{TM} = \frac{K_1[H^+]}{[H^+]^2+K_1[H^+]+K_1K_2}$$

$$\alpha_2 = \frac{[A^{2-}]}{v}\frac{v}{TM} = \frac{K_1K_2]}{[H^+]^2+K_1[H^+]+K_1K_2}$$

You can set up such a calculation in MS Excel, with calculations of α_0, α_1, α_2, at a range of different pH values ($[H^+]$ concentrations), for a given K_1 and K_2, and plot the speciation against pH.

More details are described by King *et al.* (1990) J. Chem. Educ. 67 (11), p. 932; DOI: 10.1021/ed067p932

Box 2. Example 1 for multiprotic chemicals: Carbonic acid

Let's try carbonic acid (bicarbonate, or H_2CO_3) as an example first. H_2CO_3 is the product of carbon dioxide dissolved in water. In pure water/seawater the hydration equilibrium constant $K_h = [H_2CO_3]/[CO_2] \approx 1.7\times10^{-3}$ / $\approx 1.2\times10^{-3}$ respectively, indicating that only 0.1% of dissolved CO_2 equilibrates to H_2CO_3. The dissolved concentration of CO_2 depends on atmospheric CO_2 levels according to the air-water distribution coefficient (Henry constant $k_H = pCO_2/[CO_2]$= 29.76 atm/(mol/L)). Because of the relevance of CO_2 in e.g. ocean acidification, and gas exchange in our lungs, it is interesting to see how H_2CO_3 speciates depending on pH. As in the formula H_nA, n= 2 for H_2CO_3.

$$v = \frac{([H_2CO_3])}{[H^+]^2},\ so\ \frac{([H_2CO_3])}{v} = [H^+]^2$$

With pK_1^*=6.5 (in equilibrium with atmospheric CO_2) and pK_2=10.33, $K_1=10^{-6.5}$ and $K_2=10^{-10.3}$.

At pH 7, $[H^+]=10^{-7}$, so at pH 7 with these dissociation constants

$$\alpha_0 = \frac{([H_2CO_3])}{v}\frac{v}{TM} = \frac{(10^{-7})^2}{(10^{-7})^2+(10^{-6.5})(10^{-7})+(10^{-6.5})(10^{-10.3})} = 0.24,\ so\ f_{H_2CO_3} = 24\%$$

$$\alpha_1 = \frac{[HCO_3^-]}{v}\frac{v}{TM} = \frac{(10^{-6.5})(10^{-7})}{(10^{-7})^2+(10^{-6.5})(10^{-7})+(10^{-6.5})(10^{-10.3})} = 0.76,\ so\ f_{HCO_3^-} = 76\%$$

$$\alpha_2 = \frac{[CO_3^{2-}]}{v}\frac{v}{TM} = \frac{(10^{-6.5})(10^{-10.3})}{(10^{-7})^2+(10^{-6.5})(10^{-7})+(10^{-6.5})(10^{-10.3})} = 0.00038,\ so f_{CO_3^{2-}} \sim 0\%$$

or from a series of Excel calculations at different pH:

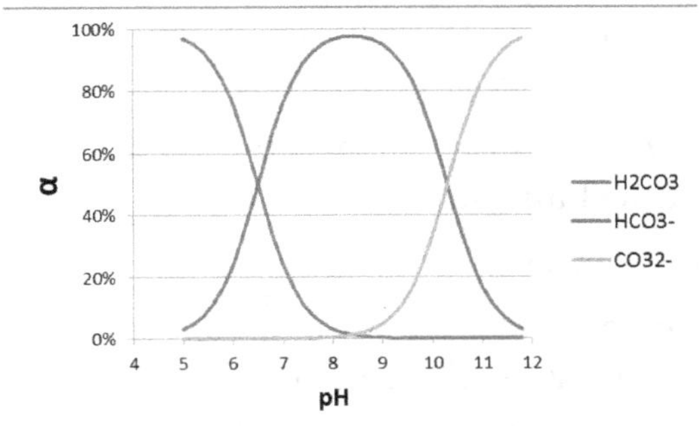

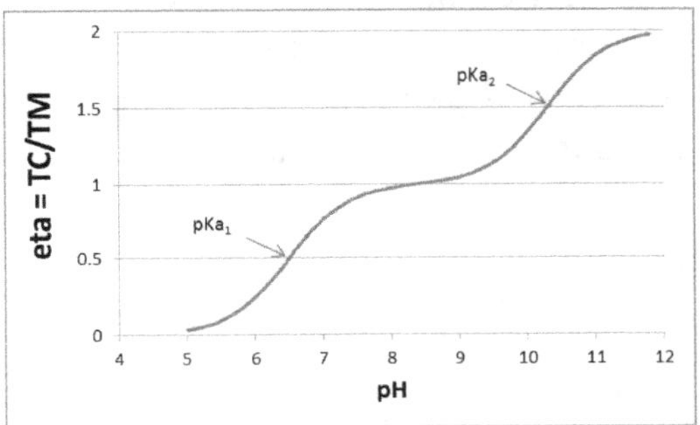

you can copy/paste the following cells into cell A1 of a new Excel sheet, and extend the range of pH:

K1	3.16E-07
K2	5.01E-11
pH	5
[H+]	=10^-B3
a0 = H_2CO_3	=(B4*B4)/((B4*B4)+(B4*B1)+(B1*B2))
a1 = HCO_{3-}	=(B4*B1)/((B4*B4)+(B4*B1)+(B1*B2))
a2 = CO_3^{2-}	=(B1*B2)/((B4*B4)+(B4*B1)+(B1*B2))
ETA	=((B4*B1)+2*(B1*B2))/((B4*B4)+(B4*B1)+(B1*B2))

Box 3. Example 2 for multiprotic chemicals: Zwitterions

Many organic pH buffers are zwitterionic chemicals, that contain both an acidic and a basic group. Norman Good and colleagues described a set of 20 of such buffers for biochemical and biological research (see for example www.interchim.fr/ft/0/062000.pdf, or www.applichem.com/fileadmin/Broschueren/BioBuffer.pdf). Examples are MES, MOPS, HEPPS (Figure A). These buffers are selected to:

- have a buffering pKa in the range of pH6-8 where most biochemical tests are performed;

- be readily soluble in water,
- be stable in test solutions, so resistant to (non)enzymatic degradation, not forming precipitates with salts
- ideally be impermeable to cell membranes so that they don't accumulate or reach active intracellular sites
- readily available, reasonably cheap.

The zwitterionic chemicals with sulfate groups are actually always have the sulfate group charged, making it highly soluble and impermeable to cell membranes, while the amine group protonates between pH6-10, depending on neighbouring functional groups. The speciation of the amine groups in MES and MOPS simply follows the single pKa calculation of equations 1-5. In HEPPS, either of the two amines is protonated, the second pKa is 3, so the doubly charged molecules only occurs at much lower pH, but can still be used as a buffer.

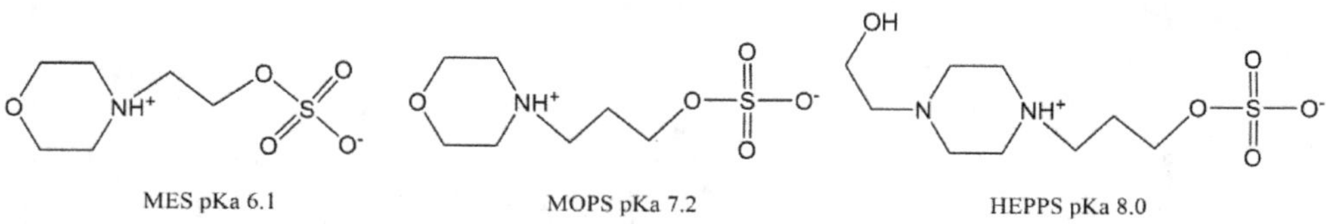

Figure A. MES, MOPS and HEPPS in charged form.

A zwitterionic chemical with two apparent pKa values relatively close is p-amino-benzoic acid. If chemicals have not one ionisable group, but N ionisable groups that speciation in a relevant pH range, than the amount of possible species is 2N. So the zwitterion p-amino benzoic acid has 4 species, each with a separate pK_a (pH where both species are present in equal concentrations).

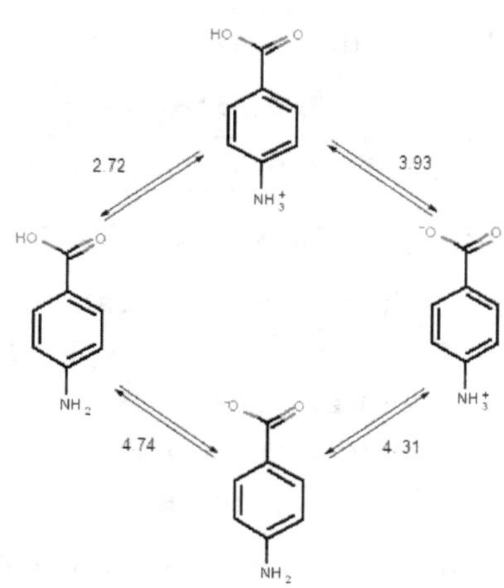

Let's formulate the benzyl group in p-amino benzoic acid as X, the neutral amine base as B, and the neutral carboxylic acid as AH, so that the fully neutral species is BXAH.

Compared to the carboxylic acid, we now have under the most acidic conditions (BHXAH)+, as α0α0, the neutral species BXAH and the zwitterionic intermediate (BHXA)0, as α1α1, and anionic species at most alkaline conditions (BXA)-, as α2α2. The calculation of the fraction of each species can be calculated according to similar rules as for carbonic acid if the two dissociation constants are known (K1 = 2.4, K2=4.88).

However, this does not inform us on the ratio between the zwitterionic form and the fully neutral form. To do this, the speciation constants of the 4 microspecies are required.

[BH+XAH] ↔ [BXAH] + [H+] for which the pk1 is calculated to be 2.72

K1=10^-2,72 = [BXAH]*[H+]/ [BH+XAH]

[BH+XAH]*10^-2.72 = [BXAH]*[H+]

which rearranges to [BXAH]= 10^-2.72 *[BH+XAH] /[H+]

[BH+XAH] ↔ [BH+XA-] + [H+] for which the pk2 is calculated to be 3.93

K2=10^-3.93 = [BH+XA-]*[H+] / [BH+XAH]

[BH+XAH]*10^-3.93 = [BH+XA-]*[H+]

which rearranges to [BH+XA-] =10^-3.93*[BH+XAH]/[H+]

[BXAH] ↔ [BXA-] + [H+] for which the pk3 is calculated to be 4.74

K3= 10^-4,74 = ([BXA-] * [H+]) / [BXAH]

[BXAH]*10^-4,74 = [BXA-] * [H+])

[BH+XA-] ↔ [BXA-] + [H+] for which the pk4 is calculated to be 4.31

K4=10^-4,31 = [BXA-]*[H+] / [BH+XA-]

[BH+XA-]*10^-4,31 = [BXA-]*[H+]

So the ratio between zwitterionic form [BH+XA-] and neutral form [BXAH] equals to:

[BH+XA-] / [BXAH] = 10^-pK2 / 10^-pK1

[BH+XA-] / [BXAH] = 10^-3.93 / 10^-2.72 = 0.06: so only 6% zwitterionic vs 94% neutral species.

The macro pK1 and pK2 are then calculated as:

K1 = ([BXAH] *[H+] + [BH+XA-]*[H+])/ [BH+AXH]

K1 = [BXAH] *[H+]/[BH+XAH] + [BH+XA-]*[H+] / [BH+XAH]

$$K1 = 10^{-pK1} + 10^{-pK2}$$

$$K1 = 10^{-2.72} + 10^{-3.93} = 10^{-2.69}$$

$$K1 = 10^{-2.69}$$

$$K2 = ([BXA^-]) * [H^+] / ([BH^+XA^-] + [BXAH])$$

$$1/K2 = [BH^+XA^-]/([BXA^-]) * [H^+] + [BXAH]/([BXA^-]) * [H^+]$$

$$1/K2 = 1/10^{-pK3} + 1/10^{-pK4}$$

$$1/K2 = 1/10^{-4,31} + 1/10^{-4,74} = 1/10^{-4.87}$$

$$K2 = 10^{-4.87}$$

2.2.7. Complex mixtures/UVCBs

Introduction/Type of Substances

Beside substances that consist of a single chemical structure, there are also substances produced that contain multiple constituents each having its unique molecular structure. In general, three types of substances can be identified: 1) mono-constituent substances, 2) multi-constituent substances, and 3) UVCBs that are substances of Unknown or Variable composition, Complex reaction products or Biological materials (ECHA, 2017a). Mono-constituent substances contain one main constituent that makes up at least 80% of the substance (Figure 1A), whereas multi-constituent substances consist of several main constituents that are present at a concentration between 10% and 80% (Figure 1B). Potential other constituents within these substances are considered impurities (Figure 1A&B). These first two substance categories are sometimes also described as well-defined substances, as the composition is (or can be) well characterized. However, this section will specifically focus on the third category, the UVCB substances. UVCBs contain many different constituents of which some can be (partially) unknown and/or the exact composition can be variable or difficult to predict (Figure 1C).

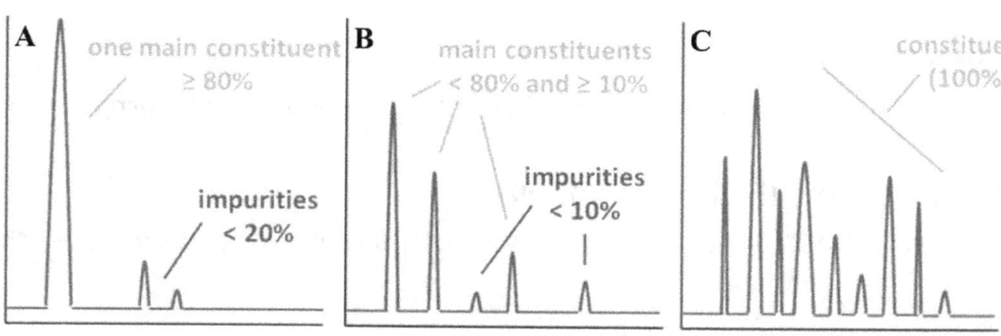

Figure 1. Overview of the three main types of substances. A) mono-constituent substance. B) multi-constituent substance. C) UVCB substance. Pictures are derived from ECHA What is a substance? - ECHA (europa.eu).

Principally, none of the constituents in a UVCB are considered as impurities. Although different terms are used to define UVCBs / complex chemical substances within various regulatory frameworks (Salvito et al., 2020), the term 'UVCB' will be used throughout this section to represent these various denominations.

Types and naming of UVCB substances

Several types of UVCB substances can be defined, including UVCBs that are synthesized or derived/refined from biological or mineral sources (ECHA, 2017a). Common types of UVCBs include:

- Extracts from plant or animal products (e.g. 'Lavender, Lavandula hybrida, ext.' [CAS 91722-69-9; EC 294-470-6]).
- Reaction products that are formed during a chemical reaction (e.g. 'Reaction product of 1,3,4-thiadiazolidine-2,5-dithione, formaldehyde and phenol, heptyl derivs.' [EC 939-460-0]).
- Products that are derived from industrial processes (e.g. 'Naphtha (petroleum), catalytic reformed' [CAS 68955-35-1; EC 273-271-8]).

In general, the name of a UVCB substance is a combination of its source (e.g. name of the species for biological sources, or name of the starting material for non-biological sources) and the used process(es) (e.g. extraction, fractionation, etc.) (ECHA, 2017a). In addition, for some UVCB categories, specific nomenclature systems are developed that can also include a description of the general composition or characteristics (e.g. physicochemical properties, like boiling range).

A specific nomenclature system is for instance developed for hydrocarbon solvents and oleochemicals, in which the nomenclature is based on the chemical composition (e.g. 'Hydrocarbons, C9-C11, n-alkanes, isoalkanes, cyclics, < 2% aromatics [CAS 64742-48-9; EC 919-857-5]') (OECD, 2014; OECD, 2015).

Challenges in the risk and hazard evaluation of UVCBs

It is difficult to fully characterize the chemical composition of UVCBs as they can contain a relative large number of constituents. Generally, it is technically challenging or impossible to identify, and thus to test, all individual constituents present in a UVCB.

As a consequence, a significant fraction is often defined as 'unknown' or is only specified in general terms (ECHA, 2017a,b). Nevertheless, specific information down to the individual constituent level could be relevant for risk/hazard assessment as some constituents can already cause effects at low concentrations (see chapter 6). In addition to an 'unknown' fraction, UVCBs can also have a variable composition. The composition may for instance depend on fluctuations in the manufacturing processes and/or source material, including spatial and temporal variations.

Although this variability may not affect the functionality of the UVCB substance, it could influence and warrant a new hazard assessment (ECHA, 2017a,b; Salvito et al., 2020). Obviously, these key characteristics of UVCBs (i.e. the compositional complexity), complicate their risk and hazard assessment.

As it is not possible to identify, isolate and assess all individual constituents, alternative assessment approaches are being developed to evaluate UVCBs, including whole-substance and constituent based approaches. Within whole-substance based approaches, the UVCB is used as test-item. Testing of the whole substance might be relevant when the UVCB consists of structurally very similar constituents that are expected to have comparable fate and effect properties (Figure 2).

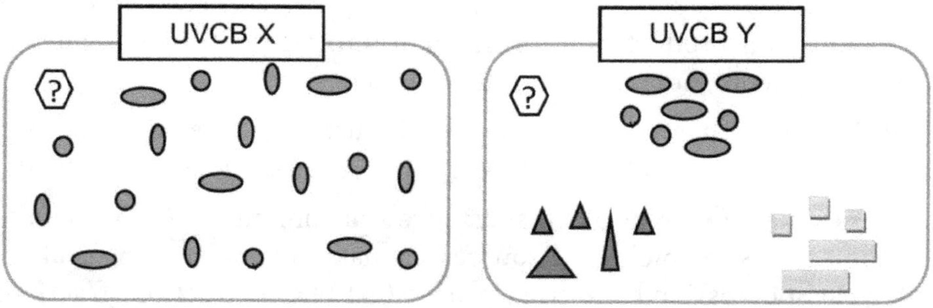

Figure 2. UVCB substances X and Y contain constituents with varying physicochemical properties (colors). For substance X a whole-substance based approach might be used, whereas for substance Y a constituent based approach is generally preferred. The different shapes represent constituents that could potentially be grouped according to other properties, such as mode of toxic action. The '?' represents an 'unknown' fraction. This figure is adopted and modified from Salvito et al. (2020).

However, when the UVCB displays a wide range of physicochemical properties a constituent based approach is generally preferred for risk/hazard assessment purposes, as the results of whole substance testing can be very difficult to interpret. For instance, the results of whole substance testing typically provide a single profile for the whole UVCB, while the fate, behavior and effects between (groups of) constituents could differ significantly. Furthermore, result interpretation might be challenging, as it could be difficult to maintain stable dosing/exposure concentrations when constituents with varying physicochemical properties are combined (e.g. due to differences in sorption, evaporation, solubility etc.).

Within constituent-based approaches, generally one or a few representative constituents are selected and evaluated. The results for these constituents are subsequently extrapolated to the other constituents, and ultimately to the UVCB. The selection of representative constituents can be based on several aspects, including in silico predictions of fate and hazard properties, the relevance or availability of the constituents and the structural variability (ECHA, 2017b; Salvito et al., 2020). To support the generation and selection of representative constituents computational methodologies are being developed (Dimitrov et al., 2015; Kutsarova et al., 2019).

One of the best described constituent based approaches is the 'fraction profiling approach', which is also known as the hydrocarbon block method (ECHA, 2017b; King, 1996). This method is specifically developed for petroleum substances (although it may also be applied to other UVCBs) and is applied in several risk assessment and PBT (Persistent, Bioaccumulative, Toxicity) assessment approaches (CONCAWE, 2016; Wassenaar and Verbruggen, 2021).

Within the hydrocarbon block method, the composition of a UVCB is conceptually divided in blocks/fractions of (structurally) similar constituents. The underlying assumption is that all constituents within a block have fairly similar properties and could be assessed as if it is 'one constituent'. More details on the hydrocarbon block method are provided in section 2.3.5 Hydrocarbons.

In general, the choice of the assessment approach is dependent on the substance and may also be dependent on the data already available as well as the stage and the general purpose of the assessment (e.g. PBT-assessment, risk assessment, etc.). In some cases, a combination of varying approaches could be considered most efficient. For instance by using a whole substance as test item in combination with analytical measurements of individual constituents over time.

References/Useful links:

CONCAWE. (2016). PETRORISK Version 7.0 Public (Manual). https://www.concawe.eu/reach/petrorisk/.

Dimitrov, S., Georgieva, D.G., Pavlov, T.S., Karakolev, Y.H., Karamertzanis, P.G., Rasenberg, M., Mekenyan, O.G. (2015). UVCB substances: methodology for structural description and application to fate and hazard assessment. Environmental Toxicology and Chemistry 11, 2450-2462. https://doi.org/10.1002/etc.3100.

ECHA (2021). What is a substance. https://echa.europa.eu/support/substance-identification/what-is-a-substance

ECHA (2017a). Guidance for identification and naming of substances under REACH and CLP. European Chemicals Agency, Helsinki.

ECHA (2017b). Guidance on Information Requirements and Chemical Safety Assessment. Chapter R.11: PBT/vPvB assessment. European Chemicals Agency, Helsinki.

King, D.J., Lyne, R.L., Girling, A., Peterson, D.R., Stephenson, R., Short, D. (1996). Environmental risk assessment of petroleum substances: the hydrocarbon block method. CONCAWE report no. 96/52. Brussels.

Kutsarova, S.S., Yordanova, D.G., Karakolev, Y.H., Stoeva, S., Comber, M., Hughes, C.B., Vaiopoulou, E., Dimitrov, S.D., Mekenyan, O.G. (2019). UVCB substances II: Development of an endpoint-nonspecific procedure for selection of computationally generated representative constituents. Environmental Toxicology and Chemistry 38, 682-694. https://doi.org/10.1002/etc.4358.

OECD. (2015). OECD guidance for characterising hydrocarbon solvents for assessment purposes. Series on Testing and Assessment, No. 230. ENV/JM/MONO (2015)52. Organization for Economic Cooperation and Development, Paris.

OECD. (2014). OECD guidance for characterising oleochemical substances for assessment purposes. Series on Testing and Assessment, No. 193. ENV/JM/MONO(2014)6. Organization for Economic Cooperation and Development, Paris.

Salvito, D., Fernandez, M., Jenner, K., Lyon, D.Y., De Knecht, J., Mayer, P., MacLeod, M., Eisenreich, K., Leonards, P., Cesnaitis, R., León-Paumen, M., Embry, M., Déglin, S.E. (2020). Improving the environmental risk assessment of substances of unknown or variable composition, complex reaction products, or biological materials. Environmental Toxicology and Chemistry 39, 2097-2108. https://doi.org/10.1002/etc.4846.

Wassenaar, P.N.H., Verbruggen, E.M.J. (2021). Persistence, bioaccumulation and toxicity-assessment of petroleum UVCBs: A case study on alkylated three-ring PAHs. Chemosphere 276, 130113. https://doi.org/10.1016/j.chemosphere.2021.130113.

2.2.8. Plastics

Introduction

Since its introduction in the 1950s, the amount of plastics in the environment has increased dramatically (Figure 1). A recent study by Jambeck *et al.* (2015) estimated that 192 coastal countries generated 275 million metric tonnes of plastic waste in 2010 of which around 8 million tons of land-based plastic waste ends up in the ocean every year. By UN Environment plastic pollution is seen as one of the largest environmental threats. If waste management does not change rapidly, another 33 billion tonnes of plastic will have accumulated around the planet by 2050. (Micro) plastics is widely recognized as a serious problem in the ocean, however, plastic pollution is also seen in terrestrial and freshwater systems.

Classification by Size and Morphology

Plastics are commonly divided into macroplastics and microplastics; the latter plastic particles are <5 mm in diameter (including nanoplastics). There are several ways to classify microplastics but the following two types are often used; primary microplastics and secondary microplastics. Primary microplastics have been made intentionally, like pellets or microbeads, secondary microplastics are fragmented parts of larger objects. Microplastics show a large variety in characteristics such as size, composition, weight, shape and color. These characteristics have an influence on the behaviour in the environment, like for instance, the dispersion in water and the uptake by organisms (Figure 2). Low-density particles float on water and are therefore more prone to advection than particles with a higher density. Similarly, spheres are more likely to be taken up by organisms than fibers. The characteristics also affect the absorption of contaminants, adsorption of microbes, and potential toxicity.

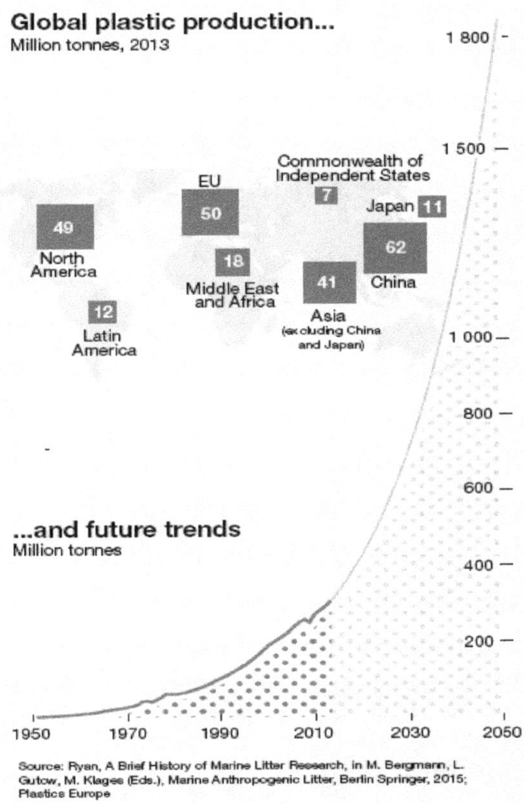

Figure 1. Global plastic production and future trends (source: http://www.grida.no/resources/6923; 2019, GRID-Arendal & Maphoto/Riccardo Pravettoni).

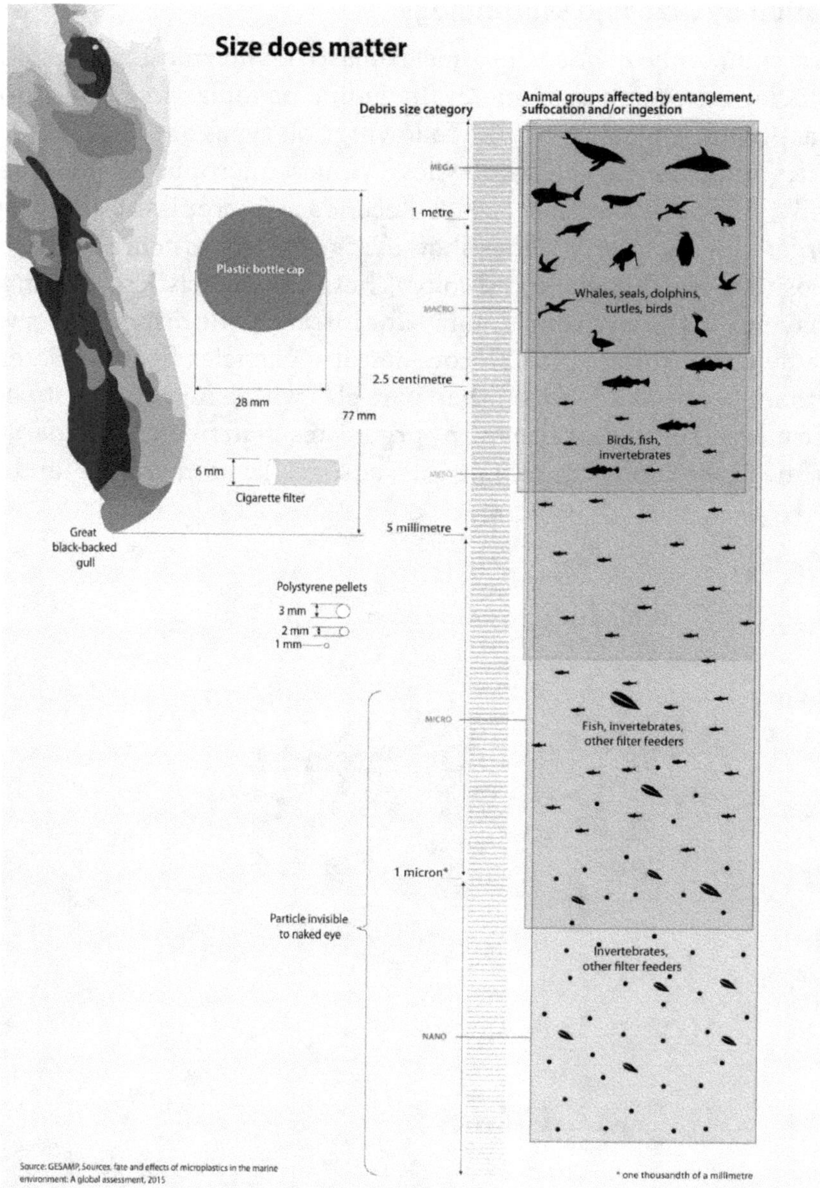

Figure 2. Marine litter comes in all sizes. Large objects may be tens of metres in length, such as pieces of wrecked vessels, lost. (Source: http://www.grida.no/resources/6924; 2019; GRID-Arendal & Maphoto/Riccardo Pravettoni).

Classification by Chemistry

Plastic is the term used to define a sub-category of the larger class of materials called polymers, usually synthesized from fossil fuels, although biomass and plastic waste can also be used as feedstock. Polymers are very large molecules that have characteristically long chain-like molecular architecture. There are many different types of plastics but the market is dominated by 6 classes of polymers: polyethylene (PE, high and low density), polypropylene (PP), polyvinyl chloride (PVC), polystyrene (PS, including expanded EPS), polyurethane (PUR) and polyethylene terephthalate (PET) (figure 3). In order to make materials flexible, transparent, durable, less flammable and long-lived, additives to polymers are used such as flame retardants (e.g. polybrominated diphenyl ethers), and plasticisers (e.g. phthalates). Some of these substances are known to be toxic to marine organisms and to humans.

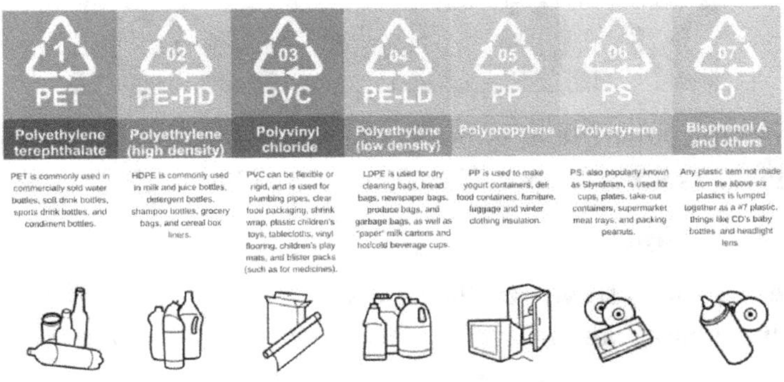

Figure 3. Different types of plastics (source https://isustainrecycling.com/plastics-recycling/)

Biopolymers/ bioplastics

There is a lot of discussion on bioplastics as degradable plastics that these may still persist for a long time under marine conditions. Please watch this video by dr. Peter Kershaw.

Plastic Degradation

Degradation of plastics takes place as soon as the plastic loses its original integrity and properties. There is a faster breaking up phase (degradation into microparticles) and a much slower mineralization phase (polymer chains being degraded to carbon dioxide). The degradation rate of plastics is determined by its polymer type, additive composition and environmental factors. Many commonly-used polymers are extremely resistant to biodegradation. Although plastics degrade in natural environments it is argued that no polymer can be efficiently biodegraded in a landfill site. Plastics in aquatic environments can be subject to in-situ degradation, e.g. by photodegradation or mechanical fragmentation but are in general very durable. As a result, plastics that are present in our oceans will degrade at a very slow pace, (Figure 4). So the majority of plastics produced today will persist in the environment for decades and probably for centuries, if not millennia.

Figure 4. "How long until it is gone" : the time required to degrade different materials. (source https://futurism.com/plastic-decomposition)

Plastics in the environment

- **Sources and Pathways:** Plastics are found in terrestrial, freshwater, estuarine, coastal and marine environments, and even in very remote areas of the world and the deep-sea. Sources and pathways of marine litter are diverse and exact quantities and routes are not fully known. But there is a surge in interest to determine the exact quantities and types of plastic litter and pathways in the environment and most of the plastic in our oceans originates from land-based sources (Figure 5) but also from sea-based sources. Most PE and PP is used in (single-use) packaging products that have a short lifetime and end up soon as waste.

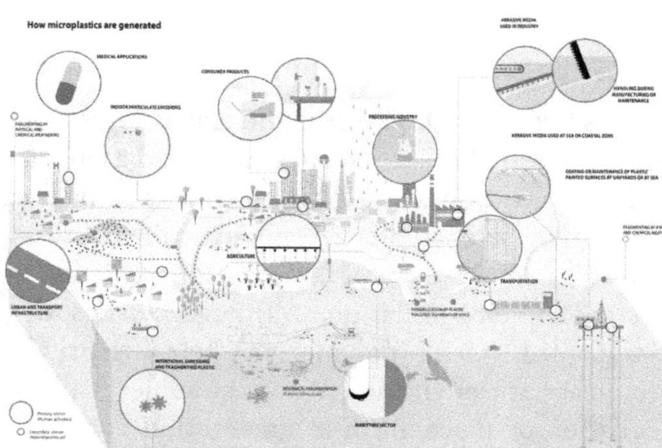

Figure 5. Overview of the major sources of primary microplastics and the generation of secondary microplastics (Source: http://www.grida.no/resources/6929; 2019; GRID-Arendal & Maphoto/Riccardo Pravettoni).

- **Effects:** As plastics have become widespread and ubiquitous in the environment, they are present in a diversity of habitats and can impact organisms at different levels of biological organization, possibly leading to population, community and ecosystem effects. Entanglement is one of the most obvious and dramatic physical impacts of macroplastics, as it often leads to acute and chronic injury or death. In particular the higher taxa (mammals, reptiles, birds and fish) are affected, and it may be critical for the success of several endangered species. Because of similar size characteristics to food, plastics are both intentionally and unintentionally ingested by a wide range of species, such as invertebrates, fish, birds and mammals. Ingestion of the non-nutritional plastics can cause damage and/or obstruction of the digestive tract and may lead to decreased foraging due to false feelings of satiation, resulting in reduced energy reserves.

Microplastics and in particular nanoparticles that are small enough to be taken up and translocated into tissues, cells and body fluids can cause cellular toxicity and pathological changes due to particle toxicity. In addition, there are also chemical risks involved as plastics can be a source of hazardous chemicals. These chemicals can be part of the plastic itself (i.e. monomers and additives) and/ or chemicals that are absorbed from the environment into the matrix or such as lead, cadmium, mercury, persistent organic pollutants (POPs) like PAHs, PCBs and dioxins. However, as this process depends on the fugacity gradients, there is a lot of uncertainty about the extent that transfer of pollutants does occur in the environment. Actually, when taking all exposure pathways into account, the transfer from (micro)plastics seems to be a minor pathway.

Watch the video on the research of Inneke Hantoro.

Finally, marine plastics may act as floating habitats for invasive species, including harmful algal blooms and pathogens, leading to spreading beyond their natural dispersal range and creating the risk of disrupting ecosystems of sensitive habitats.

2.2.9. Nanomaterials

Introduction

Engineering at the nanoscale (i.e. 10-9 m) brings the promise of radical technological development. Due to their unique properties, engineered nanomaterials (ENMs) have gained interest from industry and entered the global market. Potentials ascribed to nanotechnology are amongst all: stronger materials, more efficient carriers of energy, cleaner and more compact materials that allow for small yet complex products. Currently, nanomaterials are used in numerous products, although exact numbers are lacking. In 2014, the market was estimated to contain more than 13,000 nano-based products (Garner and Keller, 2014). There is a wide variety of products containing nanomaterials, ranging from sunscreens and paint, to textiles, medicines, electronics covering many sectors (Figure 1).

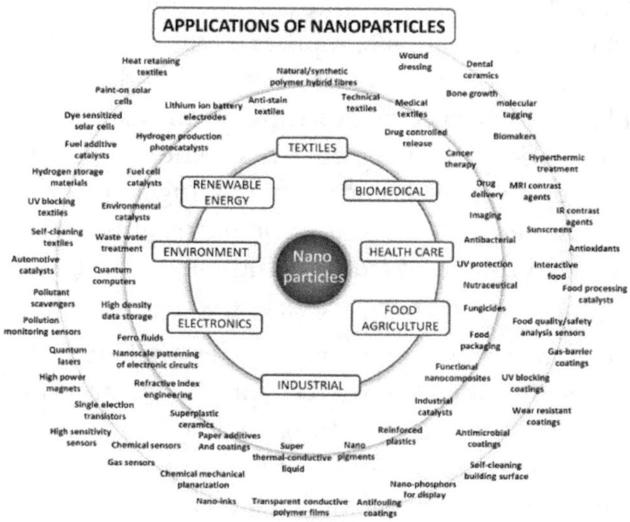

Figure 1. Applications in different sectors where engineered nanomaterials are used (source: http://www.enteknomaterials.com/wp-content/uploads/2016/08/nano-malzemeler-5.jpg).

Nanomaterial

The European Commission in 2011 adopted a new definition of 'nanomaterial' reading 'a natural, incidental or manufactured material containing particles, in an unbound state or as an agglomerate or as an aggregate and where, for 50 % or more of the particles in the number size distribution, one or more external dimensions is in the size range 1 nm-100 nm'.

Nanomaterials occur naturally, think of minuscule small fine dust, colloids in the water column, volcanic ash, carbon black and colloids known as ocean spray. In paints the features of the colloids are used to obtain the pigment colors. From the year 2000 on, an exponential growth was seen in their synthesis due to the advanced technologies and imaging techniques needed to work on a nano-scale. First generation nanotechnologies (before 2005) generally refers to nanotechnology already on the market, either as individual nanomaterials, or as nanoparticles incorporated into other materials, such as films or composites. Surface engineering has opened the doors to the development of second and third generation ENMs. Second generation nanotechnologies (2005-2010) are characterised by nanoscale elements that serve as the functional structure, such as electronics featuring individual nanowires. From 2010 onward there has been more research and development of third generation nanotechnologies, which are characterised by their multi-scale architecture (i.e. involving macro-, meso-, micro- and nano-scales together) and three-dimensionality, for applications like biosensors or drug-delivery technologies modelled on biological templates. Self-assembling bottom-up techniques have been widely developed at industrial scale, to create, manipulate and integrate nanophases into more complex nanomaterials with new or improved technological features. Post 2015, the fourth generation ENMs are anticipated to utilise 'molecular manufacturing': achieving multi-functionality and control of function at a molecular level. Nowadays, virtually any material can be made on the nanoscale.

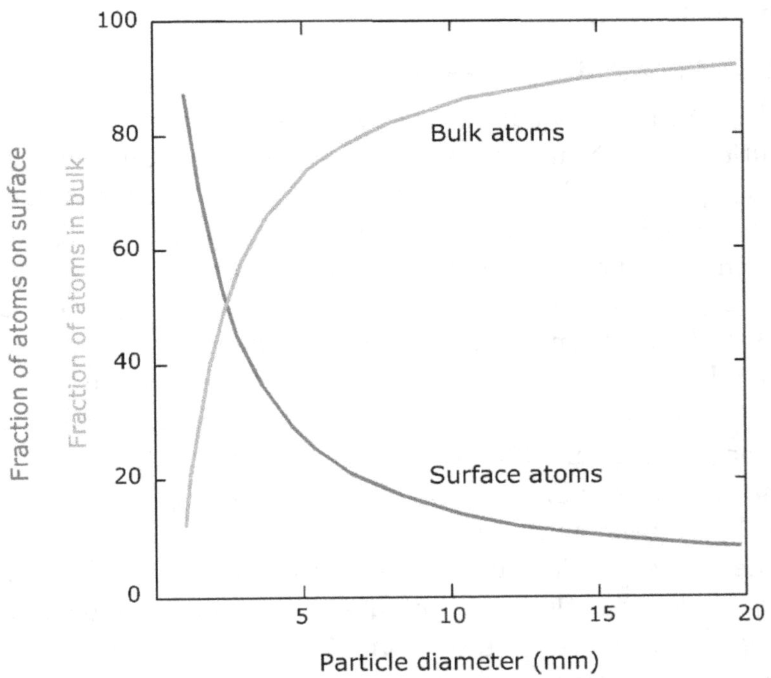

Figure 2. Relationship between particle diameter and the fraction of atoms at the surface. Drawn by Wilma Ijzerman.

Size does matter

Nanoscale materials have far larger surface areas than larger objects with similar masses. A simple thought experiment shows why nanoparticles have phenomenally high surface areas.

A solid cube of a material 1 cm on a side has 6 cm^2 of surface area, about equal to one side of half a stick of gum. When the 1 cm3 is filled with micrometer-sized cubes — a trillion (1012) of them, each with a surface area of 6 square micrometers — the total surface area amounts to 6 m^2.. As surface area per mass of a material increases, a greater proportion of the material comes into contact with surrounding materials (Figure 2). Small particles also give that there is a high proportion of surface atoms, high surface energy, spatial confinement and reduced imperfections (Figure 2). It results in the fact that ENMs are having an enlarged reactivity. compared to larger "bulk" materials. For instance, ENMs have the potency to transfer concentrated medication across the cell membranes of targeted tissues. By engineering nanomaterials, these properties can be harnessed to make valuable new products or processes.

ENMs are often designed to accomplish a particular purpose, taking advantage of the fact that materials at the nanoscale have different properties than their larger-scale counterparts.

ENMs and Environmental Processes

ENMs are described as a population of particles, and quantified by the particle size distribution (PSD). Nonetheless often a single value (e.g. average ± standard deviation) is reported and not the full PSD. When the particles are suspended in an exposure medium, the size distribution of the NPs is changing over time. After being emitted into aquatic environments, NPs are subject to a series of environmental processes. These processes include dissolution and aggregation (see Figure 3) and subsequent sedimentation. It is known that the behavior and fate of NPs are highly dependent on the water chemistry. In particular, environmental parameters like pH, concentration and type of salts (especially divalent cations) and natural organic matter (NOM) can strongly influence the behaviour of NPs in the environment. For example, pH can affect the aggregation and dissolution of metallic NPs by influencing the surface potential of the NPs (von der Kammer et al., 2010). The divalent cations Ca^{2+} and Mg^{2+} are able to efficiently compress the electrical double-layer of NPs and consequently enhance homo-aggregation and hetero-aggregation of NP (see Figure 2) and the cations will be related to bridging the electrostatic interactions. In surface water, aggregation processes most often lead to sedimentation and sometimes to floating aggregates (depending on the density).

Coating of ENMs will change the dynamics of these processes. As a result of these nano-specific features, ENMs form a suspension which is different from chemicals that dissolve and form a solution. These ENMs in suspension then follow different environmental fate and behaviour compared to solutions. For this reason the way the dosage of ENPs should be expressed is highly debated within the nano-safety community. Should this be on a mass basis, as is the case for molecules of conventional chemicals (e.g. mg/L), or is the particle number the preferred dose metric such as in colloidal science (e.g. number of particles/L, or relative surface-volume ratio) or multi-mixed dosimetry expression. How to express the dose for nanomaterials is a quest still debated within the scientific community (Verschoor et al., 2019).

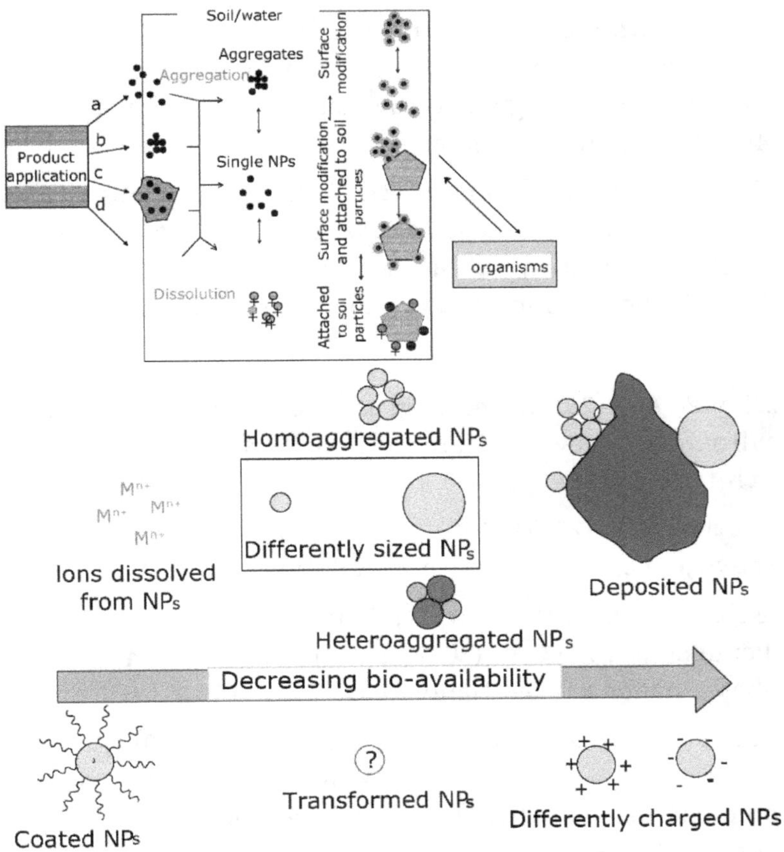

Figure 3. Top: Schematic illustration how metallic nanoparticles (NPs) behave in an exposure matrix (redrawn from Nowak & Bucheli, 2007). Bottom: Aggregation is divided into two type: homo-aggregation (i.e., aggregation between nanoparticles) and hetero-aggregation (i.e., aggregation of nanoparticle and biomass). Drawn by Wilma IJzerman.

Classification of NMs

Although we learned from the text above that changing the form of a nanomaterial can produce a material with new properties (i.e. a new nanomaterial); often a group of materials developed is named after the main chemical component of the ENMs (e.g. nanoTiO_2) that is available in different (nano)forms. Approaches to group ENMs have been presented below:

- **Classification by dimensionality / shape / morphology**: Shape-based classification is related to defining nanomaterials, and has been synopsized in the ISO terminology.
- **Classification by composition / chemistry:** This approach groups nanomaterials based on their chemical properties.
- **Classification by complexity / functionality:** The nanomaterials that are in routine use in products currently are likely to be displaced by nanomaterials designed to have multiple functionalities, so called 2nd-4th generation nano-materials.
- **Classification by biointerface:** A proposal relates to the hypothesis that nanomaterials acquire a biological identity upon contact with biofluids and living entities. Systems biology approaches will help identify the key impacts and nanoparticle interaction networks.

References

Garner, K.L., Keller, A.A. (2014). Emerging patterns for engineered nanomaterials in the environment: a review of fate and toxicity studies. Journal of Nanoparticle Research 16: 2503.

Nowack, B., Bucheli, T.D. (2007). Occurrence, behavior and effects of nanoparticles in the environment. Environmental Pollution 150, 5-22.

Verschoor, A.J., Harper, S., Delmaar, C.J.E., Park, M.V.D.Z., Sips, A.J.A.M., Vijver, M.G., Peijnenburg, W.J.G.M. (2019). Systematic selection of a dose metric for metal-based nanoparticles. NanoImpact 13, 70-75.

Von der Kammer, F., Ottofuelling, S., Hofmann, T. (2010). Assessment of the physico-chemical behavior of titanium dioxide nanoparticles in aquatic environments using multi-dimensional parameter testing. Environmental Pollution 158, 3472-3481.

2.3. POLLUTANTS WITH SPECIFIC USE

2.3.1. Crop Protection Products

Introduction

Crop protection products are used in agriculture. The principle target of agriculture is the provision of food. For this purpose, agriculture aims to reduce the competition by other (non-crop) plants and the loss of crop due to herbivores or diseases. An important tool to achieve this is the use of chemicals, such as crop protection products (CPP). Accordingly, CPPs are intentionally introduced into the environment and represent one of the largest sources of xenobiotic chemicals in the environment. These chemicals are by definition effective against the target organism, often already at fairly low doses, but may also be toxic to non-target organisms including humans. The use of pesticides, also named Crop Protection Products (CPP) or often also Plant Protection Products (PPP, the latter term may be misleading for herbicides which are intended to reduce all but the crop plants), is therefore strictly regulated in most countries. The main pesticides used in the largest volumes world-wide are herbicide all s, insecticides, and fungicides. As shown in Table 1, pesticides are used against a large number of diseases and plagues.

Table 1. Classification of pesticides according to what they are supposed to control

Pesticide type	Target
acaricides	against mites and spiders (incl. miticides)
algicides	against algae
althelmintics (vermicides)	against parasites
antibiotics	against bacteria and viruses (incl. bactericides)
bactericides	against bacteria
fungicides	against fungi
herbicides	against weeds
insecticides	against insects
miticides	against mites
molluscicides	against slugs and snails
nematicides	against nematodes
plant growth regulators	retard or accelerate the growth of plants
repellents	drive pests (e.g. insects, birds) away
rodenticides	against rodents

Formulations

A pesticidal product usually consists of one or more active substances, that are brought onto the market in a commercial formulation (spray powder, granulate, liquid product etc.). The formulation is used to facilitate practical handling and application of the chemical, but also to enhance its effect or its safety of use. The active substance may, for instance, be a solid chemical, while application requires it to be sprayed. Or the active substance degrades fast under the influence of sunlight and therefore has to be encapsulated. One of the most used types of formulation is a concentrated emulsion, which may be sprayed directly after dilution with water. In this formulation, the active substance is dissolved in an oily matrix and a detergent is added as emulsifier to make the oil miscible with water. In this way, the active substance becomes quickly available after spraying. In so-called slow-release formulations, the active substance is encapsulated in permeable microcapsules, from which it is slowly released. Another component of a formulation can be a synergist, which increases the efficacy of the active substance, for instance by blocking enzymes that metabolize the active substance. Here an overview of main formulation constituents:

- **Solvents:** to ease handling and application of the active substances, they are usually dissolved. For a highly water soluble compound this solvent may just be water; however, most compounds have low water solubility and they are thus dissolved in organic solvents.
- **Emulsifier, detergents, dispersants:** used to provide a homogeneous mixture of the active substance in the aqueous spray solution.
- **Carrier:** solid formulations, such as wettable granules (WG). All wettable powder (WP) formulations often use inert materials such as clay (kaolinite) as carrier.
- **Wetting agent:** they help providing a homogeneous film on the plant surface.
- **Adjuvant:** may help to increase uptake of the active substance into the plant.
- **Minor constituents:**
    - **Antifreeze agent:** to keep the formulation stable also in cold storage conditions.
    - **Antifoam agent:** some of the formulants may result in foaming during application, which is not wanted.
    - **Preservative, biocide:** to prevent biological degradation of the active substance or the formulants during storage.
    - There are numerous additional specific additional constituents for specific purposes such as colours, detterents, stickers, etc

Four types of nomenclature are used in case of pesticides:

- The trade name, e.g. Calypso®, which is given by the manufacturer. The same active substance is often sold under more than one different trade names (accordingly, the use of trade names only is not a sufficient description of the test substance in scientific literature).
- The code name, which is the "common" name of the active substance. Calypso® 480 SC, for example, is a concentrated suspension containing 480 g/L of the active substance thiacloprid.

- The chemical name of the active substance. Thiacloprid is [3-[(6-chloropyridin-3-yl)methyl]-1,3-thiazolidin-2-ylidene] cyanamide.
- The name of the chemical group to which the active substance belongs, in case of thiacloprid: neonicotinoids.

Chemical Classes

Pesticides represent quite a number of different groups of chemicals. Pesticides include inorganic chemicals (like copper used as a fungicide), organic synthetic chemicals, and biologicals (organic natural compounds). Pesticides from the same chemical group may be used against different pest organisms, like the organotin compounds (see below). Some chemicals have a broad mode of action: many soil disinfectants, such as metam-sodium, kill nematodes, fungi, soil insects and weeds. Other pesticides are more selective, like neonicotinoids acting only on insects, or very selective, like the insect-growth regulator fenoxycarb, which is used against leaf-rollers without affecting its natural enemies. Selectivity of a pesticide also indicates to what extent non-target species may be affected upon its application (side-effects). Integrated pest management (IPM) aims at an as sustainable as possible crop protection system by combining biological agents (predators of the pest organism) using chemicals having a selective mode of action. Such systems are nowadays receiving increasing interest in different agricultural crops.

Some groups of pesticides that were used or still are widely used are presented in more detail. Their modes of action are discussed in Chapter 4.

Chlorinated Hydrocarbons

Best known representative of this group is DDT (dichloro diphenyl trichloroethane; Figure 1), which was discovered in 1939 by the Swiss entomologist Paul Hermann Müller and seemed to be an ideal pesticide: it was effective, cheap and easy to produce and remained active for a long period of time. As a remedy against Malaria and other insect borne diseases, it has saved millions of human lives. However, the high persistency of DDT, its strong bioaccumulation and its effects on bird populations have triggered the search for alternatives and its ban in most Western countries. But in some developing countries, because of a lack of suitable alternatives for an effective control of malaria, DDT is still in use to kill malaria mosquitos.

Other representatives of chlorinated hydrocarbons are lindane, also called gamma-hexachlorocyclohexane (Figure 1), and the cyclodienes that include the "drins" (aldrin, dieldrin, endrin, See Section 2.1) and endosulfan (Figure 1). Because of their high persistence and bioaccumulative potential, most organochlorinated pesticides have been banned.

Volatile halogenated hydrocarbons were often used as soil disinfectant. These compounds were injected into the soil, and acted as a nematicide but also killed fungi, soil insects and weeds. An example is 1,3-dichloropropene (Figure 1).

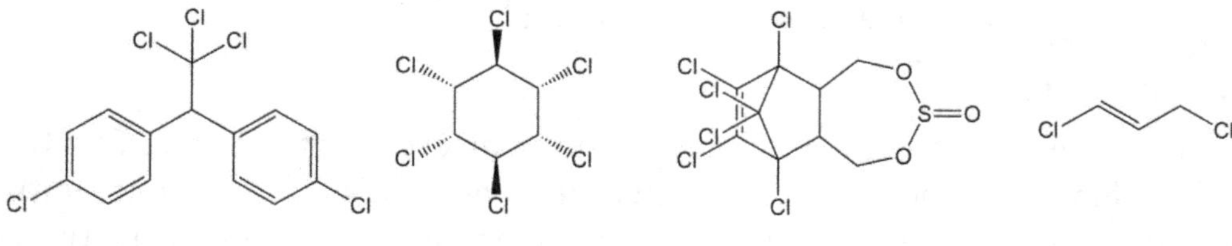

Figure 1. Chemical structures of four different organochlorinated pesticides widely used in the past, from left to right: DDT: 1,1'-(2,2,2-trichloroethane-1,1-diyl)bis(4-chlorobenzene), lindane: gamma-1,2,3,4,5,6-hexachlorocyclohexane, endosulfan: 6,7,8,9,10,10-hexachloro-1,5,5a,6,9,9a-hexahydro- 6,9-methano-2,4,3-benzodioxathiepine-3-oxide, and 1,3-dichloropropene. (Source: Steven Droge)

Organophosphates

Organophosphates are esters of phosphoric acid and constitute important biological molecules such as nucleic acids (DNA) or ATP. Within the contents of pesticides this refers mainly to a group of organophosphate molecules which interfere with acetylcholinesterase. Nerve gases, produced for chemical warfare (e.g., Sarin), also belong to the organophosphates. They are much less persistent and were therefore introduced as alternatives for the chlorinated hydrocarbons. The common molecular structure of organophosphates is a tri-ester of phosphate, phosphonate, phosphorthionate, phosphorthiolate, phosphordithionate or phosphoramidate (Figure 2). With two of the three ester bonds, a methyl- or ethyl- group is bound to the P atom, while the third ester bond binds the rest group or "leaving group".

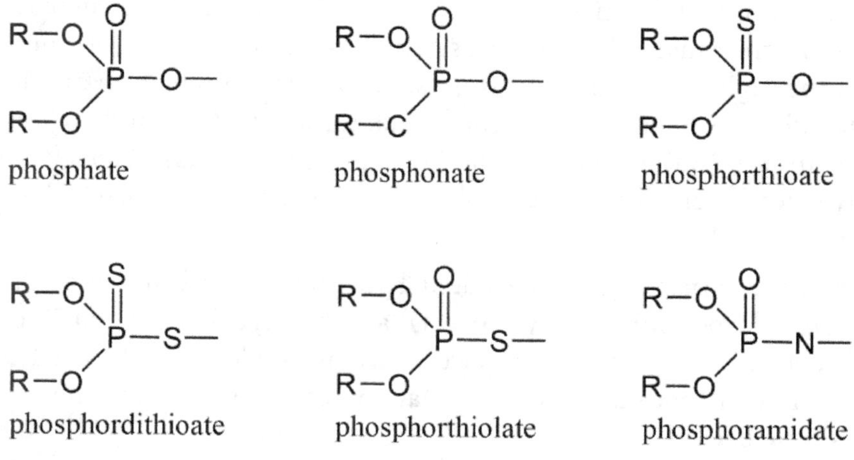

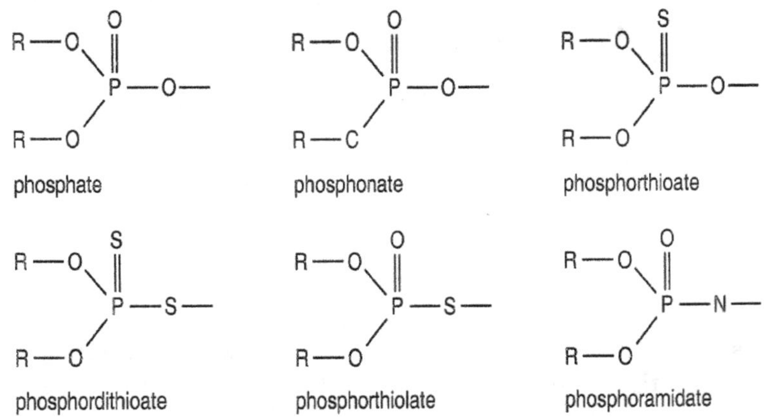

Figure 2: Chemical structure of organophosphates. R = methyl of ethyl group. (Source: Steven Droge)

Dependent on the identity of the latter group, three sub-groups may be distinguished:

- Aliphatic organophosphates, including malathion (Figure 3) and a number of systemic chemicals.
- Phenyl-organophosphates, which are more stable than the aliphatic ones but also less soluble in water, like parathion (no longer allowed in Europe; Figure 3).
- Heterocyclic organophosphates, including chemicals with an aromatic ring containing a nitrogen atom like chlorpyrifos (Figure 3).

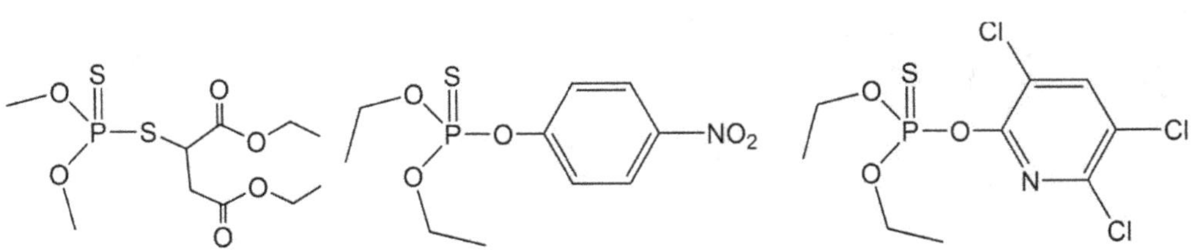

Figure 3: Malathion: diethyl 2-[(dimethoxyphosphorothioyl)sulfanyl]butanedioate (left), parathion: O,O-diethyl-O-4-nitrophenyl-phosphorthioate (middle), and chlorpyrifos: O,O-diethyl O-3,5,6-trichloropyridin-2-yl phosphorothioate (right). (Source: Steven Droge)

Carbamates

Where organophosphates are derived from phosphoric acid, carbamates are derived from carbamate (Figure 4). Their mode of action is similar to that of the organophosphates. The use of older representatives of this group, like aldicarb, carbaryl, carbofuran and propoxur, is no longer allowed in Europe, but diethofencarb (Figure 4), oxamyl and methomyl are still in use.

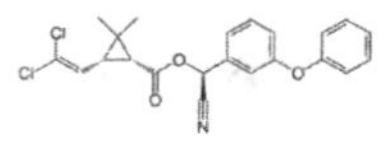

Figure 4: Basic structure of carbamates (left) and diethofencarb: isopropyl 3,4-diethoxycarbanilate (right). (Source: Steven Droge)

Pyrethroids

A number of modern pesticides are derived from natural products. Pyrethroids are based on pyrethrum, a natural insecticide from flowers of the Persian ox-eyed daisy, Chrysanthemum roseum. Typical for the molecular structure of pyrethroids is the cyclopropane-carboxyl group (the triangular structure), which is connected with an aromatic group through an ester bond (Figure 5). Pyrethrum is rapidly degraded under the influence of sunlight. Synthetic pyrethroids, which are much more stable and therefore used on a large scale against many different insects, include cypermethrin (Figure 5), deltamethrin, lambda-cyhalothrin, fluvalinate and esfenvalerate.

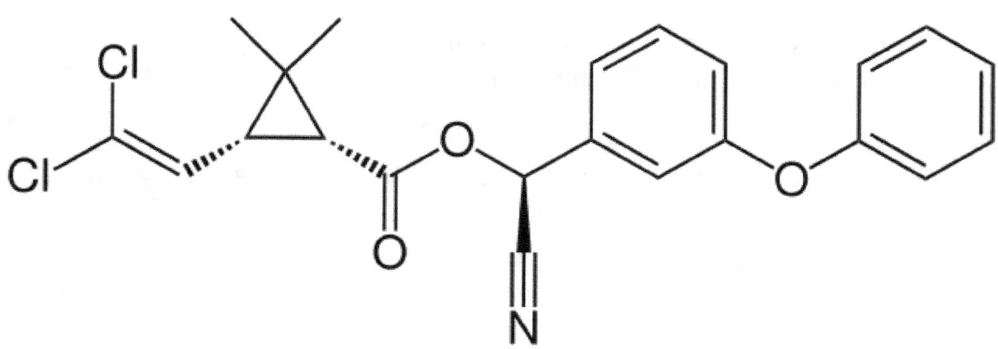

Figure 5: Cypermethrin: [cyano-(3-phenoxyphenyl)methyl]3-(2,2-dichloroethenyl)-2,2-dimethylcyclopropane-1-carboxylate. (Source: Steven Droge)

Neonicotinoids

Based on the natural compound nicotine, which acts as a natural insecticide against plant herbivores, but which was banned as an insecticide due to its high human toxicity, in the 1980s a new group of more specific insecticides has been developed, the neonicotinoids (Figure 6). Several neonicotinoids (e.g., imidacloprid, thiamethoxam) are systemic. This means that they are taken up by the plant and exert their effect from inside the plant, either on the pest organism (systemic fungicides or insecticides) or on the plant itself (systemic herbicides). The systemic neonicotinoids are widely applied as seed dressing in major crops like maize and sunflower. Other compounds are mainly used in spray applications, e.g. in fruit growing (thiacloprid, acetamiprid, etc.). Although neonicotinoids are more selective and therefore preferred over the older classes of insecticides like organophosphates, carbamates and pyrethroids, in recent years they have become under debate because of their side effects on honey bees and other pollinators.

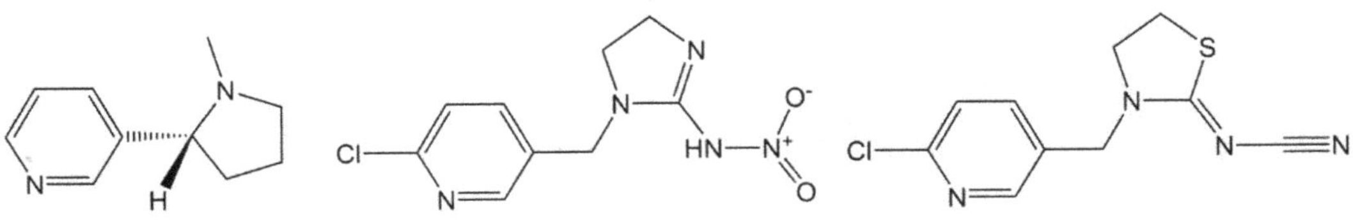

Figure 6: Nicotine: (S)-3-[1-methylpyrrolidin-2-yl]pyridine (left) and the neonicotinoid insecticides imidacloprid: N-{1-[(6-chloro-3-pyridyl)methyl]-4,5-dihydroimidazol-2-yl}nitramide (middle) and thiacloprid: {(2Z)-3-[(6-chloropyridin-3-yl)methyl]-1,3-thiazolidin-2-ylidene}cyanamide (right). (Source: Steven Droge)

Isothiocyanates

Isothiocyanates were used on a large scale as soil disinfectant against nematodes, fungi and weeds. The large number of chemicals with different chemical origin belonging to the isothiocyanates have in common that they form isothiocyanate in soil. A representative of this group is metam-sodium (Figure 7).

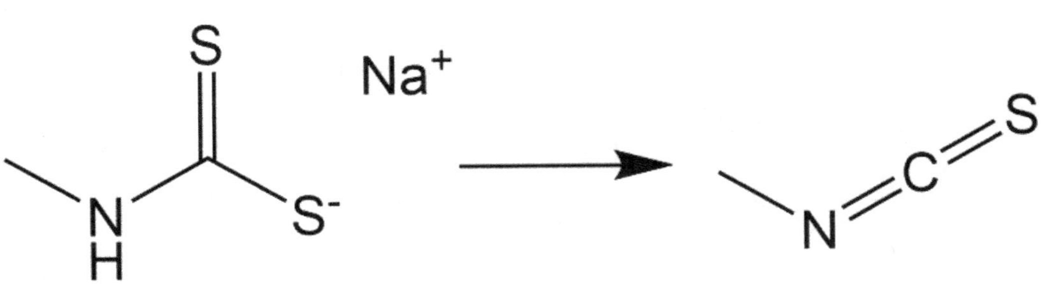

Figure 7. Metam-sodium: sodium methylaminomethanedithioate forming methyl isothiocyanate. (Source: Steven Droge)

Organotin Compounds

Fentin hydroxide (Figure 8) was used as a fungicide against *Phytophthora* (causing potato -disease). Tributyltin compounds (TBT) were used as anti-fouling agent (algicide) on ships. TBTC (tributyltin chloride) is extremely toxic to shell-fish, such as oysters, and for this reason banned in many countries. Fenbutatin-oxide was used as an acaricide against spider mites on fruit trees (tributyltin chloride).

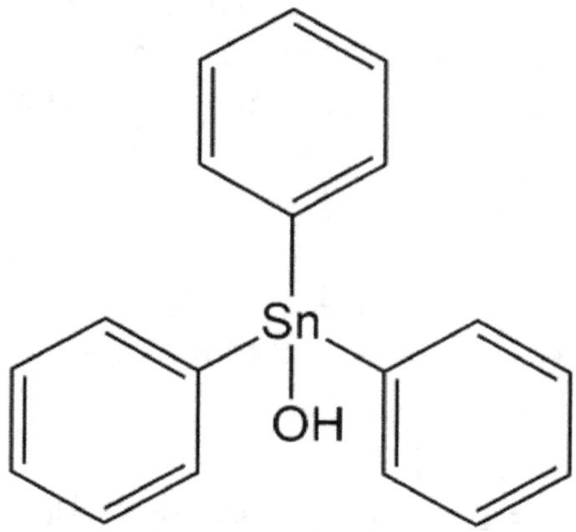

Figure 8: Fentin hydroxide: triphenyltin hydroxide. (Source: Steven Droge)

Ryanoids

Also indicated as diamide insecticides, this group includes chemically distinct synthetic compounds such as chlorantraniliprole (Figure 9), flubendiamide, and cyantraniliprole, that act on the ryanodine receptor and are used against chewing and sucking insects.

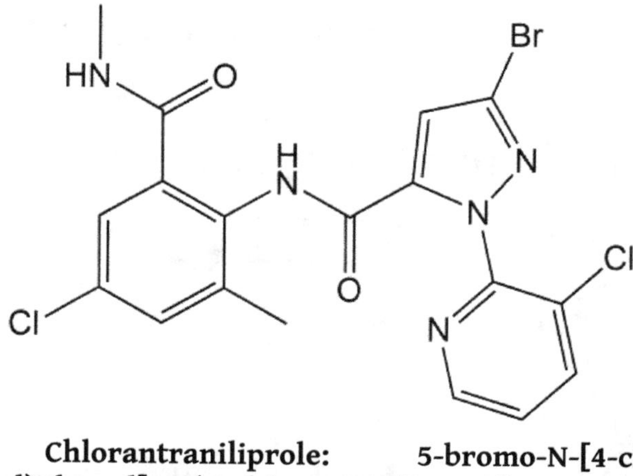

Figure 9: Chlorantraniliprole: 5-bromo-N-[4-chloro-2-methyl-6-(methylcarbamoyl)phenyl]-2-(3-chloropyridin-2-yl)pyrazole-3-carboxamide. (Source: Steven Droge)

Phenoxy Acetic Acids

Phenoxy acetic acids are systemic herbicides, exerting their action after uptake by the leaf and translocation throughout the plant. Especially plants with broad, horizontally oriented leaves are sensitive for these herbicides. 2,4-D (Figure 10) is the best known representative of this group.

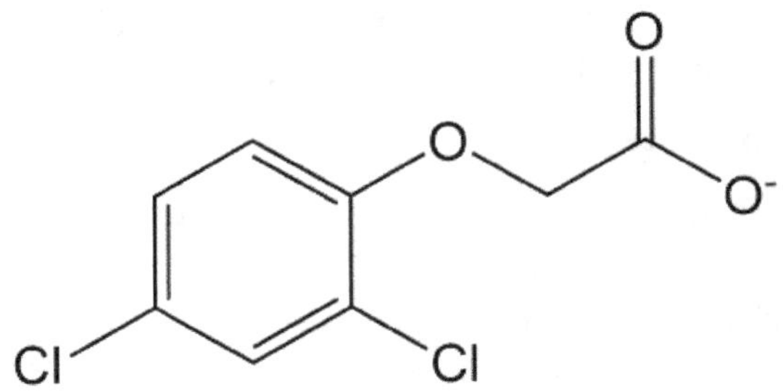

Figure 10. 2,4-D: the anionic form of 2,4-dichloro phenoxy acetic acid (pKa 2.73). (Source: Steven Droge)

Triazines

Triazines are heterocyclic nitrogen compounds, whose structure is characterized by an aromatic ring in which three carbon atoms have been replaced by nitrogen atoms. Triazines are usually applied to the soil before seed germination. The use of several compounds (atrazine, simazine) has been banned in Europe, while others like metribuzin and terbuthylazine (Figure 11) are still in use.

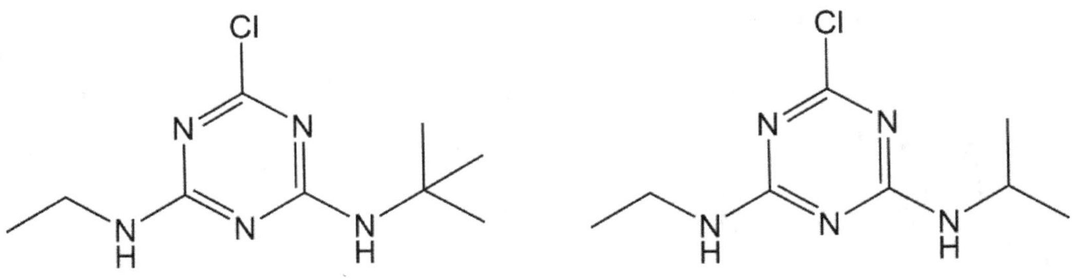

Figure 11. Terbuthylazine: N-tert-butyl-6-chloro-N'-ethyl-1,3,5-triazine-2,4-diamine (left), common replacement of the EU-banned herbicide atrazine (right). (Source: Steven Droge)

Bipiridyls

This group contains the herbicides diquat and paraquat (Figure 12) which mainly act as contact herbicides. This means they damage the plant without being taken up. In soil, they are rapidly inactivated by strong binding to soil particles. The use of paraquat is no longer allowed in Europe, but diquat is still in use.

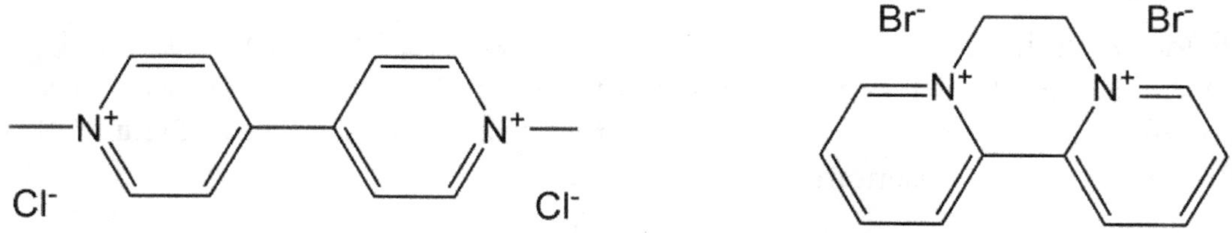

Figure 12. Paraquat: 1,1′-Dimethyl-4,4′-bipyridinium dichloride (left), and diquat: 1,1'-Ethylene-2,2'-bipyridylium dibromide (right). (Source: Steven Droge)

Glyphosate and Glufosinate

As an alternative to the above mentioned herbicides, glyphosate and later glufosinate were developed. These are systemic broad-spectrum herbicides with a relatively simple chemical structure (Figure 13). Their low toxicity to other organisms triggered pesticide producers to introduce genetically modified crops (e.g. soybean, maize, oilseed rape, and cotton) that contain incorporated genes for resistance against these broad-spectrum herbicides. This type of resistance allows the farmer to use the herbicide without damaging the crop. For this reason, environmentalist fear an unrestricted use of these herbicides, which indeed is the case especially for glyphosate (better known under the formulation name Roundup®).

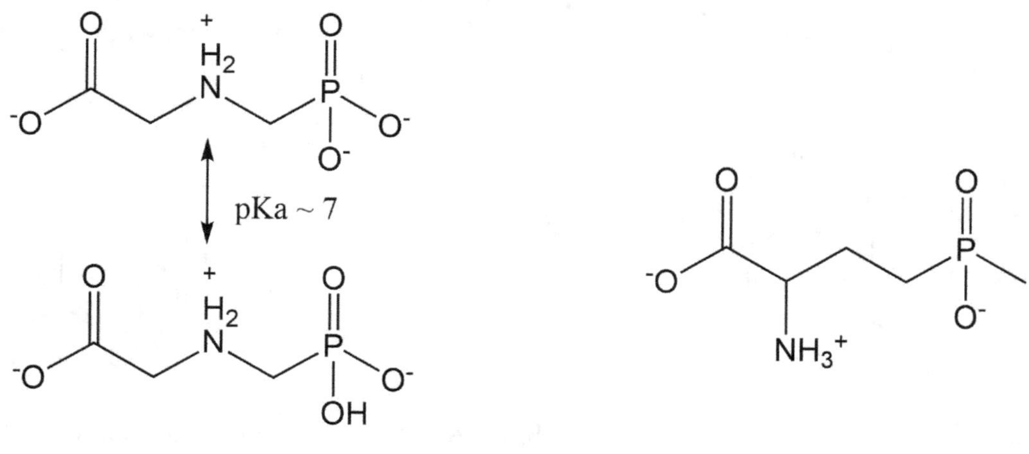

Figure 13. Glyphosate: N-(phosphonomethyl)glycine in the two species most relevant for natural pH range (left), and glufosinate: (RS)-2-Amino-4-(hydroxy(methyl)phosphonoyl)butanoic acid in the most relevant species for natural pH range (right). (Source: Steven Droge)

Triazoles

Several modern fungicides are sharing a triazole group (Figure 14). These fungicides have gained importance because of problems with the resistance of fungi against other classes of fungicides. Members of this group for instance are epoxiconazole, propiconazole and tebuconazole.

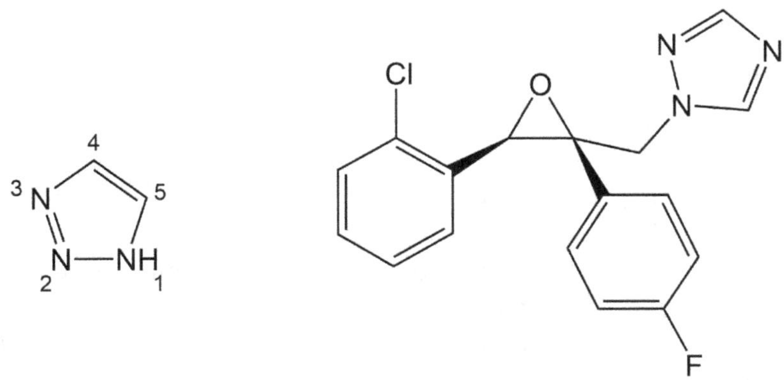

Figure 14: Triazole: 1H-1,2,3-Triazole (left), and epoxiconazole: (2RS,3SR)-1-[3-(2-chlorophenyl)-2,3-epoxy-2-(4-fluorophenyl)propyl]-1H-1,2,4-triazole (right). (Source: Steven Droge)

Biological pesticides

Biological pesticides are produced in living organisms as secondary metabolites to protect themselves against predators, herbivores, parasites or competition. They can be highly effective and act at low concentrations (high toxicity), but in contrast to some synthetic pesticides they are usually sufficiently biodegradable. Compounds like pyrethrum or strobilurin are produced within the plant or within the fungus and are thus protected against photolysis or other environmental degradation. Furthermore, the living organism can produce additional quantities of the secondary metabolite on demand. When used as a pesticide applied as a spray, however, the molecule needs to be modified to enhance its stability (for example against photolysis) to remain sufficiently active over a sufficient period of time. Accordingly, synthetic derivatives of these biological molecules are often more stable, less biodegradable. Examples are the Bt insecticide, which contains an endotoxin highly toxic to insects produced by the bacterium *Bacillus thuringiensis*, and avermectins, complex molecules synthesized by the bacterium *Streptomyces avermitilis*. Avermectins act as insecticides, acaricides and have anthelminthic properties. In nature, eight different forms of avermectin have been found. Ivermectin is a slightly modified structure that is synthesized and marketed commercially. Other compounds belonging to this group are milbemectin and emamectin.

Genetically modified plants containing a gene coding for the toxin produced by the bacterium *Bacillus thuringiensis* (or Bt) are another example of genetic modification being applied in agriculture produce insect-resistant crops.

References

EU Pesticides Database

2.3.2. Biocides

Introduction

European legislation describes a biocide as 'chemical substance or microorganism intended to destroy, deter, render harmless, or exert a controlling effect on any harmful organism by chemical or biological means'. The US Environmental Protection Agency (EPA), an independent agency of the U.S. federal government to protect the environment, defines biocides as 'a diverse group of poisonous substances including preservatives, insecticides, disinfectants and pesticides used for the control of organisms that are harmful to human or animal health or that cause damage to natural or manufactured products'. The definition by the EPA includes pesticides (Chapter 2.3.1). In the scientific and non-scientific literature, the distinction between biocides, pesticides and plant protection products is often vague.

Biocides are used all around us:

- The toothpaste that you used this morning contains biocides to preserve the toothpaste
- The water that you used to flush you mouth is prepared with biocides for disinfection
- The clothes that you are wearing are impregnated with biocides to prevent smells
- The food that you ate for breakfast might have contained biocides to preserve the food
- The construction materials around you have surface coatings that contain biocides to prevent biological degradation of the material

A biocide contains an 'active substance', which is the chemical that is toxic to its target organism, and often contain 'non-active co-substances', which could help in reaching desired product parameters, such as a viscosity, pH, colour, odour or increase its handling or effectiveness. The combination of active substances and non-active substances together makes up the 'biocidal product'. An example of a well-known biocidal product is TriChlor, which contains active substance chlorine that is used to disinfect swimming pools. Because it is impractical to store chlorine gas for the treatment of swimming pools, TriChlor tablets are added to the pool water. TriChlor is trichloroisocyanuric acid (Figure 1). When dissolved in water, the Cl atoms are replaced by H atoms, forming chlorine (Cl-) and cyanuric acid (Figure 2). The free chlorine is able to disinfect the swimming pool.

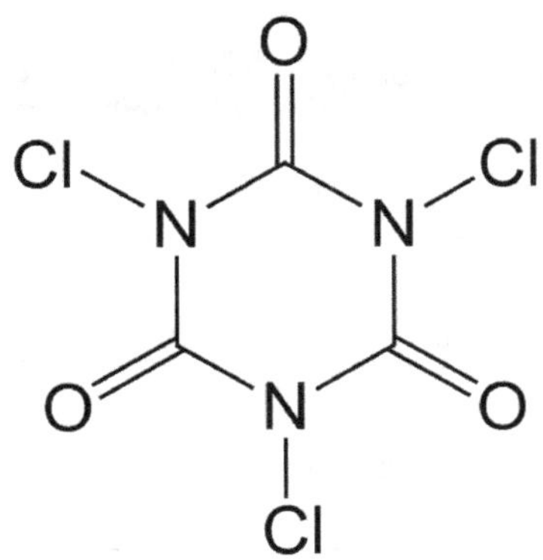

Figure 1. Trichloroisocyanuric acid (A)

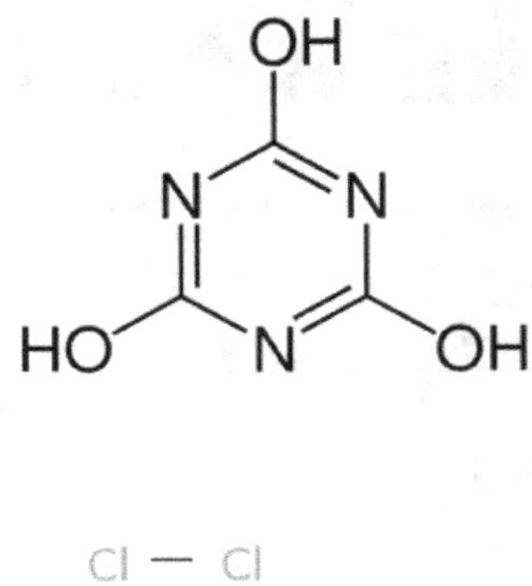

Figure 2. Cyanuric acid and chlorine (B)

A biocidal product can also contain multiple biologically active substances to enhance its effectivity, such as AQUCAR™ 742 produced by DuPont. It contains glutaraldehyde (Figure 3) and quaternary ammonium compounds (Figure 4) that have a synergistic toxic effect on microorganisms that are present in oilfields and could form biofilms in the pipelines.

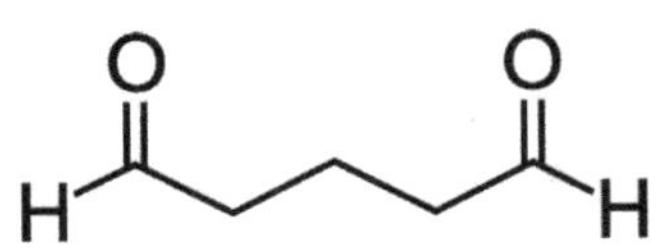

Figure 3. Glutaraldehyde

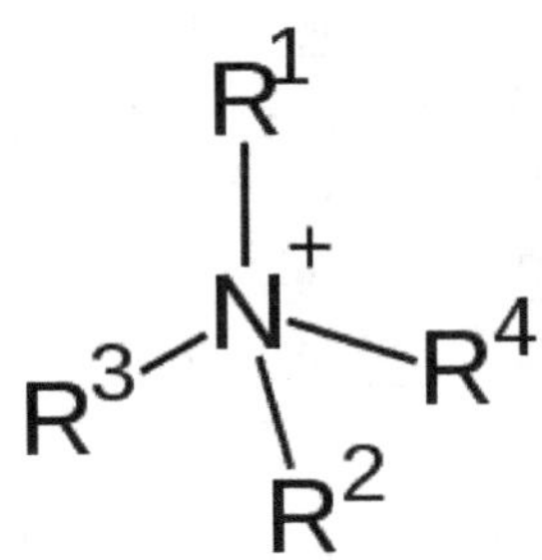

Figure 4. Quaternary ammonium compound

Product Types

The biocidal products are classified into 22 different product-types by the European Chemicals Agency (ECHA) (Table 1). It is possible that an active substance can be classified in more than one product types.

Table 1. The classification of biocides in 22 product types (www.echa.europe.eu)

Main group 1: Disinfectants and general biocidal products
Product type 1 – Human hygiene biocidal products
Product type 2 – Private area and public health area disinfectants and other biocidal products
Product-type 3 – Veterinary hygiene biocidal products
Product type 4 – Food and feed area disinfectants

Product-type 5 – Drinking water disinfectants
Main group 2: Preservatives
Product-type 6 – In-can preservatives
Product-type 7 – Film preservatives
Product-type 8 – Wood preservatives
Product-type 9 – Fibre, leather, rubber and polymerised materials preservatives
Product-type 10 – Masonry preservatives
Product-type 11 – Preservatives for liquid-cooling and processing systems
Product-type 12 – Slimicides
Product-type 13 – Metalworking-fluid preservatives
Main group 3: Pest control
Product-type 14 – Rodenticides
Product-type 15 – Avidicides
Product-type 16 – Molluscicides
Product-type 17 – Piscicides
Product-type 18 – Insecticides, acaricides and products to control other arthropods
Product-type 19 – Repellents and attractants
Product-type 20 – Control of other vertebrates
Main group 4: Other biocidal products
Product-type 21 – Antifouling products
Product-type 22 – Embalming and taxidermist fluids

Legislation

In Europe, biocides are authorised for production and use by the Biocidal Products Regulation (BPR, Regulation (EU) 528/2012) of the ECHA. The BPR 'aims to improve the functioning of the biocidal product market in the EU, while ensuring a high level of protection for humans and the environment.' (https://echa.europe.eu/legislation). This is an alternative regulatory framework than that for the plant protection products, managed by the European Food Safety Authority (EFSA). All biocidal products go through an extensive authorisation process before they are allowed on the market. The assessment of a new active starts with the evaluation of a product by the authorities of an ECHA member state, after which the ECHA Biocidal Products Committee forms an opinion. The European Commission then makes a decision to approve or reject the new active substance based on the opinion of ECHA. This approval is granted for a maximum of 10 years and needs to be renewed after it reaches the end of the registration period. The BPR has strict criteria for new active substances, and meeting the following 'exclusion criteria' will result in the new active substance not being approved:

- Carcinogens, mutagens and reprotoxic substances categories 1A or 1B according to CLP regulation
- Endocrine disruptors
- Persistent, bioaccumulative and toxic (PBT) substances
- Very persistent and very bioaccumulative (vPvB) substances

In very special cases, new active substances will be allowed on the market when meeting this exclusion criteria, if they are important for public health and public interest and there are no alternatives available. To lower the pressure on public health and the environment, there is also a candidate list for active substances to be substituted for less harmful active substances when the old active substances meet the following criteria:

- It meets one of the exclusion criteria
- It is classified as a respiratory sensitizer
- Its toxicological reference values are significantly lower than those of the majority of approved active substances for the same product-type and use
- It meets two of the criteria to be considered as PBT
- It causes concern for human or animal health and for the environment even with very restrictive risk management measures
- It contains a significant proportion of non-active isomers or impurities

The Impact of Environmental Release

The release of biocides in the environment can have huge consequences, since these products are designed to cause damage to living organisms. A classic example is the release of tributyltin from shipyards, harbours and on sailing routes from the antifouling paint on the hulls of ships (De Mora, 1996). Tributyltin was used in the antifouling paint from the 1950s on to prevent microorganisms from settling on the hulls of ships, which would increase the fuel costs and repair costs. However, the release of tributyltin from the paint resulted in a toxic effect on organisms at the bottom of the food chain, such as algae and invertebrates. Tributyltin then biomagnified in the food web, this way affecting larger predators, such as dolphins and sea otters. Eventually, tributyltin entered the diet of humans. The first legislation on the use of tributyltin for ships dates back to the 1980s, but it was not until the Rotterdam Convention of 2008 that the complete use of tributyltin as an active biocide in antifouling paints was banned. Biocides can also have an effect on the capability of the environment to deal with pollution. Microorganisms are responsible for cleaning polluted areas by using the pollutant as food-source.

McLaughlin et al. (2016) studied the effect of the release of biocide glutaraldehyde in spilled water from hydraulic fracturing on the microbial activity and found that the microbial activity was hampered by the biocide glutaraldehyde. Hence, because of the biocide, the environment was not or slower capable to return to its original state.

References

De Mora, S.J. (1996). Tributyltin: case study of an environmental contaminant, Vol. 8, Cambridge Univ. Press

McLaughlin, M.C., Borch, T., Blotevogel, J. (2016). Spills of hydraulic fracturing chemicals on agricultural topsoil: Biodegradation, sorption and co-contaminant interactions, Environmental Science & Technology 50, 6071-6078

2.3.3. Pharmaceuticals and Veterinary Pharmaceuticals

Introduction

Pharmaceuticals are used by humans (human pharmaceuticals) and administered to animals (veterinary pharmaceuticals).

The active ingredients used in human and veterinary medicine partially overlap, however, the major fraction of pharmaceutically active substances in use are restricted to human consumption. Next to that, some active ingredients are used in other applications as well, such as biocides or in plant protection products. In veterinary practice most of the applied pharmaceuticals are antibiotics and anti-parasitic agents, while in human medicine, pharmaceuticals to treat e.g. diabetes, pain, cardiovascular diseases, autoimmune disorders and neurological disorders make up a much larger portion of the pharmaceuticals in use. Worldwide pharmaceutical consumption has increased over the last century (several numbers are summarized here).

It is expected that the consumption will further increase due to a wider access to pharmaceuticals in developing countries. Additionally, demographic trends such as aging populations often seen in developed countries can also lead to increased consumption of pharmaceuticals, since older generations generally consume more pharmaceuticals than younger ones (van der Aa *et al.* 2010). The widespread and increasing use and their biological activity makes them relevant for environmental research. Pharmaceuticals are specifically designed and used for their biological effect in humans or treated animals. For that reason, we know a lot about their potential environmental effects as well as on their application and emission. Below an overview is given on the emission, occurrence and fate(modeling) of pharmaceuticals in the environment.

Pharmaceuticals in the environment

Pharmaceuticals can enter the environment through various routes. Figure 1 gives an overview of the major emission routes of pharmaceuticals to the environment.

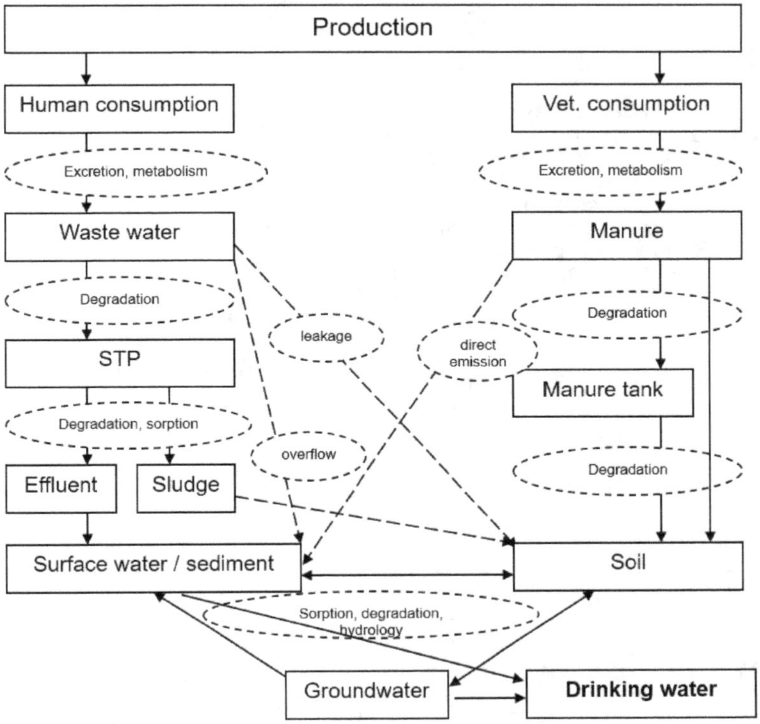

Figure 1. Pharmaceutical emissions routes to the (aqueous) environment. STP = sewage treatment plant (adapted from Schmitt et al., 2017).

Pharmaceutical are produced, transported to users (humans and/or animals), used by humans or animals. After use, the active ingredients are partially metabolized and both parent compounds and metabolites can be excreted by the users via urine and feces. For humans, the major routes are transport to wastewater treatment plants, septic tanks or directly emission to soil or surface water. For animals, and especially livestock, manure contains a major fraction of the pharmaceuticals that are excreted. These pharmaceuticals end up in the environment when animals are grazing outside or when centrally collected manure is applied as fertilizer on arable land. The treatment and further application of communal wastewater and manure varies between countries and regions. Subsequently, emissions can also vary leading to different compositions and concentrations of pharmaceuticals and metabolites in the environment. In Figure 2 concentration ranges of pharmaceuticals and some of their transformation products in the Meuse river and some tributaries are shown.

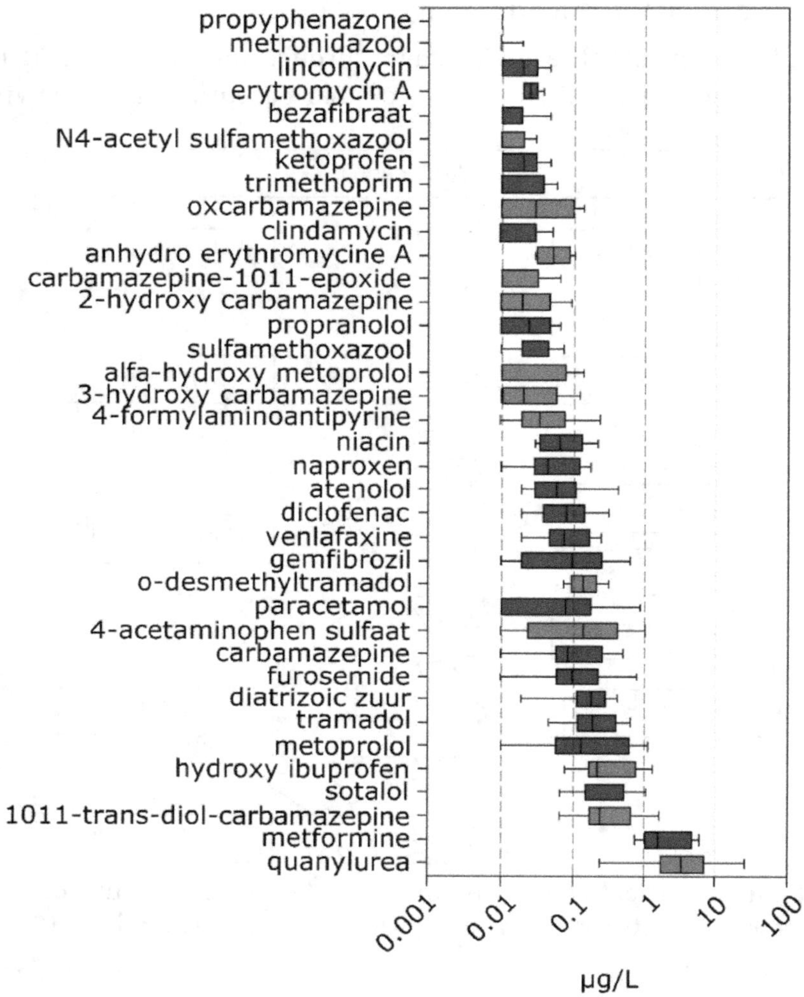

Figure 2. Pharmaceutical concentrations of pharmaceuticals in the River Meuse and some of its tributaries (adapted from Ter Laak et al. 2014). Parent pharmaceuticals are plotted in bleu, transformation products in red. Drawn by Wilma IJzerman.

Properties of Pharmaceuticals and their Behavior and Fate in the Environment

Pharmaceuticals in use are developed for a wide array of diseases and therapeutic treatments. The chemical structures of these substances are therefore also very diverse, considering their size, structural presence of specific atoms, and physicochemical properties such as their hydrophobicity, aqueous solubility and ionization under environmentally relevant pH values, as shown for some examples in Figure 3.

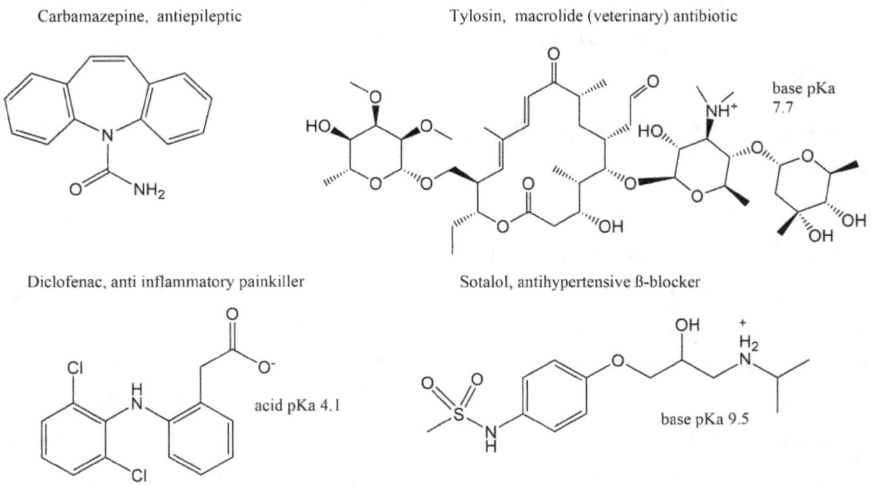

Figure 3. Examples of pharmaceuticals, illustrating the variable chemical structures. (Source: Steven Droge)

As a consequence of their structural diversity, the environmental distribution and fate of pharmaceuticals is also very variable. Nevertheless, pharmaceuticals have generally certain properties in common:

- Pharmaceuticals are designed to have a specific biological activity that determines their pharmacological application.
- Most pharmaceuticals are rather robust against metabolism of their users (humans or animals) in order to reach stable therapeutic levels inside the user.
- Pharmaceuticals are often relatively soluble in water, since their therapeutic application requires absorption and distribution to reach specific targets sites in living organisms. Aqueous solutions such as blood are often the internal transport medium. Less soluble pharmaceuticals are metabolized to allow renal excretion, leading to soluble metabolites.

These three generic properties also make them of environmental relevance since:

- Pharmaceuticals are likely biologically active in non-target organisms, thereby disturbing their behavior, metabolism or other functions.
- Pharmaceuticals are rather persistent in water treatment or the environment
- Pharmaceuticals and/or metabolites have rather high aqueous solubility which makes them mobile and as a consequence they may end up in (ground)water.

- Continuous use leads to continuous emission and subsequently continuous presence in environmental waters, this is called 'pseudo persistence'.

Occurrence and Modelling of Human and Veterinary Pharmaceuticals in the Environment

Pharmaceuticals in the environment have been studied since the 1990s. Most studies have been performed in surface waters, but wastewater (effluents), groundwater, drinking water, manure, soil and sediments were also studied. Pharmaceuticals have been observed in all these matrices in concentrations generally varying from µg/L to sub ng/L levels (Aus der Beek et al., 2016, Monteiro and Boxall, 2010). Various studies have related environmental loads and related concentrations to human consumption data. Basically such mass balance or studies that relate consumption in catchments of streams, lakes or rivers to environmental concentrations work according the following principle:

Modelling Pharmaceuticals in the Environment

The consumption of human pharmaceuticals, is relatively well documented and data are (publicly) available. Hence, based on consumption using several assumptions, environmental concentrations of pharmaceuticals can be related to consumption. This prediction works best for the most persistent pharmaceuticals, as these pharmaceuticals are hardly affected by transformation processes that can be variable as a result of environmental conditions. When loss factors become larger, they generally also become more variable, through seasonal variations in use as well as variation in loss during wastewater treatment and loss processes in the receiving rivers. This makes the loads and concentrations of more degradable pharmaceuticals more difficult to predict (Ter Laak et al., 2010).

Loads in a particular riverine system (such as a tributary of the river Meuse in the example below) can be predicted with a very simplified model. Here the pharmaceutical consumption over a selected period is multiplied by the fraction of the selected pharmaceuticals that is excreted unchanged by the human body (ranging from 0 to 1) and the fraction that is able to pass the wastewater treatment plant (WWTP) (ranging from 0 to 1):

$$Load\,River\,(\frac{kg}{day})\,Human\,Consumption(\frac{kg}{day})*fraction\,Excreted*fraction\,Passing\,WWTP$$

When this is related to actual measured concentrations and loads calculated from these numbers, the correlation between predicted and measured loads can be plotted. Various studies have shown that environmental loads can be predicted within a factor of 3 for most commonly observed pharmaceuticals (see e.g., Ter Laak et al., 2010, 2014).

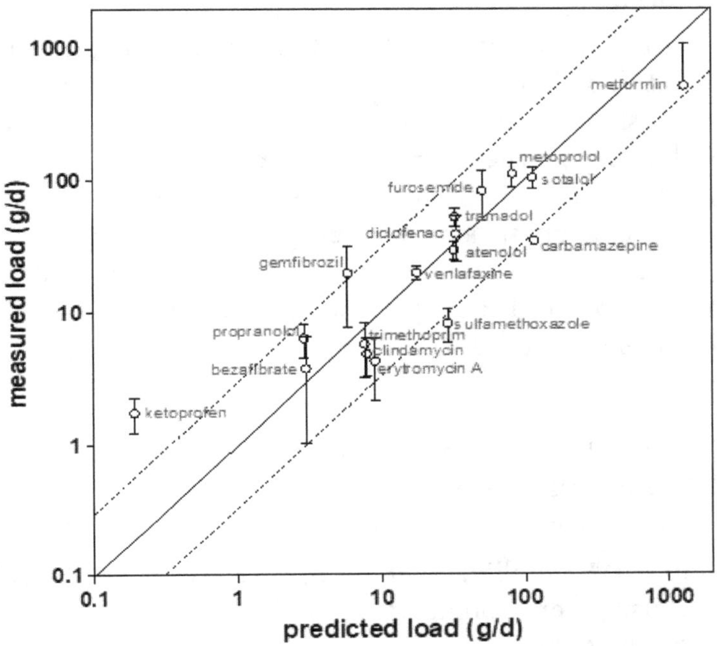

Figure 4. Measured versus predicted loads in a tributary of the Meuse river (adapted from Ter Laak et al., 2014)

For veterinary pharmaceuticals this so called 'immision-emission balancing' is more difficult for a number of reasons (see e.g., Boxall et al., 2003):

- First, veterinary pharmaceuticals are applied in different quantities and using different application routes in different (live-stock) animals.
- Second, animal excrements can be burned, stored for later use as fertilizes or directly emitted when animals are kept outside, leading to different emissions per route.
- Finally, the fate of pharmaceuticals associated with animal excrements to soil, groundwater and surface water is variable and poorly understood.

In a way the emissions and fate of veterinary pharmaceuticals is similar to emissions of pesticides used in agriculture. However, the understanding on loads entering the system and the fate related to the various emission routes and emissions in combination with a complex matrix (urine, feces manure) is more limited (Guo et al., 2016). As a consequence, environmental fate studies of veterinary pharmaceuticals often describe specific cases, or cover laboratory studies to unravel specific aspects of the environmental fate of these pharmaceuticals (Kaczala and Blum, 2016, Kümmerer, 2009).

Concluding Remarks

Pharmaceuticals are commonly found in the environmental compartments such as surface water, soil, sediment and groundwater (Williams et al., 2016). Pharmaceuticals consist of a single or multiple active ingredients that have a specific biological activity. The therapeutic application and pharmacological mechanisms provide valuable information to evaluate the environmental hazard of these chemicals. Their physicochemical properties are of more relevance for the assessment of the environmental fate and exposure. The occurrence in the environment and the biological activity of this group of contaminants makes them relevant in environmental science.

References

Aus der Beek, T., Weber, F., Bergmann, A., Hickmann, S., Ebert, I., Hein, A., Küster, A. (2016). Pharmaceuticals in the environment-Global occurrences and perspectives. Environmental Toxicology and Chemistry 35, 823-835.

Boxall, A.B.A., Kolpin, D.W., Halling-Soerensen, B., Tolls, J. (2003). Are veterinary medicines causing environmental risks? Environmental Science and Technology 37, 286A-293A.

Guo, X.Y., Hao, L.J., Qiu, P.Z., Chen, R., Xu, J., Kong, X.J., Shan, Z.J., Wang, N. (2016). Pollution characteristics of 23 veterinary antibiotics in livestock manure and manure-amended soils in Jiangsu province, China. Journal of Environmental Science and Health Part B: Pesticides, Food Contaminants, and Agricultural Wastes 51, 383-392.

Kaczala, F., Blum, S.E. (2016). The occurrence of veterinary pharmaceuticals in the environment: A review. Current Analytical Chemistry 12, 169-182.

Kümmerer, K. (2009). The presence of pharmaceuticals in the environment due to human use - present knowledge and future challenges. Journal of Environmental Management 90, 2354-2366.

Monteiro, S.C., Boxall, A.B.A. (2010). Occurrence and fate of human pharmaceuticals in the environment. Reviews of Environmental Contamination and Toxicology 202, 53-154.

Schmitt, H., Duis, K., ter Laak, T.L. (2017). Development and dissemination of antibiotic resistance in the environment under environmentally relevant concentrations of antibiotics and its risk assessment - a literature study. (UBA-FB) 002408/ENG; Umweltbundesamt: Dessau-Roßlau, January 2017; p 159.

Ter Laak, T.L., Kooij, P.J.F., Tolkamp, H., Hofman, J. (2014). Different compositions of pharmaceuticals in Dutch and Belgian rivers explained by consumption patterns and treatment efficiency. Environmental Science and Pollution Research 21, 12843-12855.

Ter Laak, T.L., Van der Aa, M., Houtman, C.J., Stoks, P.G., Van Wezel, A.P. (2010). Relating environmental concentrations of pharmaceuticals to consumption: A mass balance approach for the river Rhine. Environment International 36, 403-409.

Van der Aa, N.G.F.M., Kommer, G.J., van Montfoort, J.E., Versteegh, J.F.M. (2011). Demographic projections of future pharmaceutical consumption in the Netherlands. Water Science and Technology 63, 825-832.

Williams, M., Backhaus, T., Bowe, C., Choi, K., Connors, K., Hickmann, S., Hunter, W., Kookana, R., Marfil-Vega, R., Verslycke, T. (2016). Pharmaceuticals in the environment: An introduction to the ET&C special issue. Environmental Toxicology and Chemistry 35, 763-766.

2.3.4. Drugs of abuse

Introduction

Since about little more than a decade, drugs of abuse (DOA) and their degradation products have been recognized as emerging environmental contaminants. They are among the growing number of chemicals that can be observed in the aquatic environment.

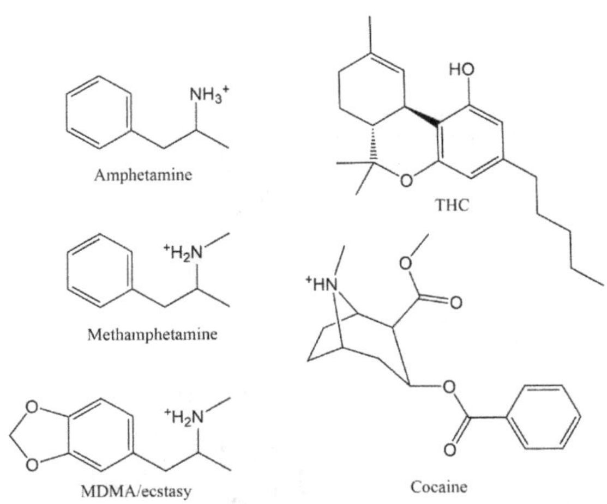

Figure 1. Chemical structures of most popular drugs of abuse, in their predominant speciation under physiological conditions (pH7.4). (Source: Steven Droge)

The residues of a major part of the chemicals used in households and daily life end up in our sewer systems. Among the many chemicals are cleaning agents and detergents, cosmetics, food additives and contaminants, pesticides, pharmaceuticals, and surely also illicit drugs. Once in the sewer, they are transported to wastewater treatment plants (WWTPs), where they may be removed by degradation or adsorption to sludge, or end up in the effluent of the plant when removal is incomplete.

The consumption of both pharmaceuticals and DOA has increased substantially over the last couple of decades as a result of several factors, including ageing of the population, medicalization of society and societal changes in life-style. As a result the loads in wastewater of drugs and their transformation products formed in the body after consumption have steadily increased.

More recently, it has been observed that chemical waste from production sites of illicit drugs is being occasionally discharged into sewer systems, thereby dramatically increasing the loads of illicit drug synthesis chemicals and end products transported to WWTPs. As WWTPs are not designed to remove drugs, a substantial fraction of the loads may end up in receiving waters and thus pose a threat to both human and ecosystem health.

Drugs of Abuse (DOA)

Europe's most commonly used illicit drugs are THC (cannabis), cocaine, MDMA (ecstasy) and amphetamines. The structure of these drugs is given in Figure 1. Other important DOA include the opioids such as heroine and fentanyl, GHB, Khat and LSD.

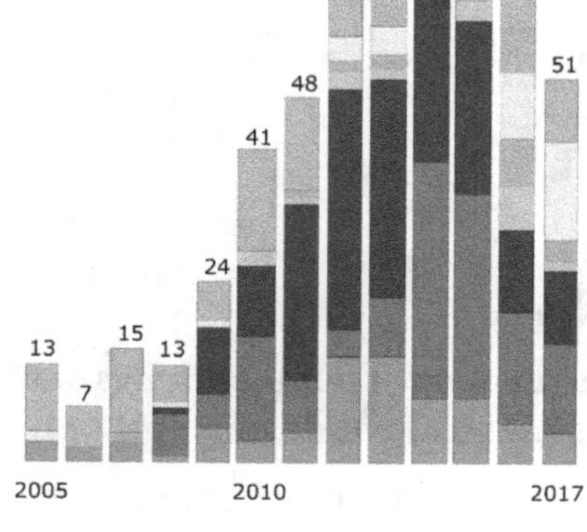

Figure 2. Number and categories of new psychoactive substances notified to the EU Early Warning System for the first time, 2005–2017. Redrawn from EMCDDA, European Drug Report 2018, Lisbon, by Wilma Ijzerman.

Drugs of abuse are controlled by legislation, in The Netherlands by the Opium Act. The Opium Act encompasses two lists of substances. List one chemicals are called hard drugs while List II chemicals are known as soft drugs. Some narcotics are also being used for medicinal purposes, e.g., ketamine, diazepines, and one of the isomers of amphetamine. New psychoactive substances (NPS), also known as designer drugs or legal highs (because they are not yet controlled as they are not listed on the Opium Act lists), are synthesized every year and become available on the market in high numbers (see Figure 2).

Wastewater-based Epidemiology

Central sewage systems collect and pool wastewater from household cleaning and personal care activities as well as excretion products resulting from human consumption and thus contain chemical information on the type and amount of substances used by the population connected to the sewer. Drugs that are consumed are metabolized in the body and subsequently excreted. Excretion products can include the intact compounds as well as the transformation products, that can be used as biomarkers. An example of the latter is benzoylecgonine, which is the major transformation product of consumed cocaine.

The collective wastewater from the sewer system carrying the load of chemicals is directed to the WWTP, and this wastewater influent can be sampled at the point where it enters the WWTP. By appropriate sampling of the influent during discrete time-intervals, e.g. 24 h, a so-called composite sample can be obtained and the concentrations of the chemicals can be determined.

The volume of influent entering the WWTP is recorded continuously. Multiplying the observed 24 h average concentration of a compound with the total 24 h volume yields the daily load of the chemical entering the WWTP. This load can be normalised to the number of people living in the sewer catchment, resulting in a load per inhabitant.

The loads of drugs in wastewater influents are usually expressed as mg.day-1.1000 inh-1. Normalised drug load data allow comparison between sewer catchments, such as shown in Figure 3. Obtaining chemical information about the population through wastewater analysis is known as Wastewater-based epidemiology, WBE (Watch the video).

While WBE was developed originally to obtain data on consumption of DOA, the methodology has been shown to have a much wider potential: in calculating the consumption of e.g., alcohol, nicotine, NPS, pharmaceuticals and doping, as well as for assessing community health indicators, such as incidence of diseases or stress biomarkers.

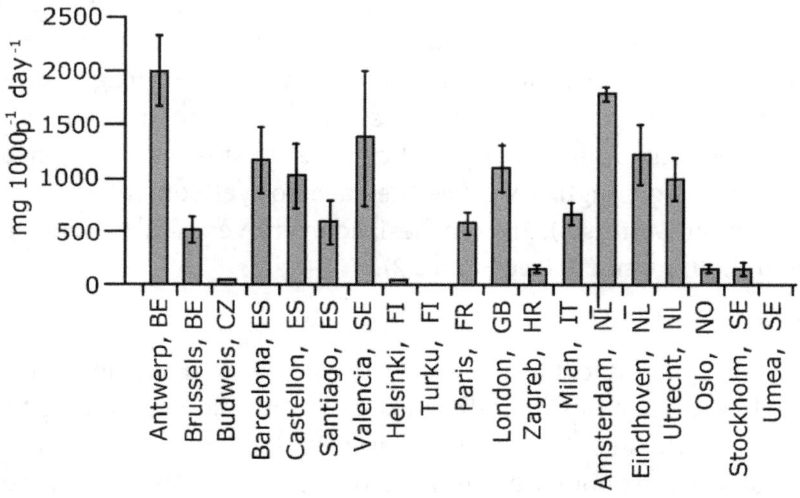

Figure 3. Consumption of cocaine in 19 European cities in 2011 calculated from chemical analysis of influent loads of benzoylecgonine, a urinary biomarker of human cocaine consumption. Redrawn from Thomas et al. (2011) by Wilma Ijzerman.

DOA and The Environment

Barring direct discharges into surface waters or terrestrial environments, the major sources of DOA to the environment are WWTP effluents. Conventional treatment in municipal WWTPs has not been specifically designed to remove pharmaceuticals or DOA. Removal rates of DOA vary widely and depend on compound properties such as persistence and polarity as well as WWTP operational conditions and process configurations. Some DOA cross WWTPs almost unhindered, thus ending op in the receiving waters. Examples of the latter are MDMA and some diazepines (see Figure 4). Despite that several studies report the presence of DOA or their transformation products in surface waters, until now there is very little information about their aquatic ecotoxicity available in the scientific literature.

Recently, chemical waste from synthetic DOA manufacturing including their precursors and synthesis byproducts have been observed to be discharged directly into sewers. In addition, containers with chemical waste from DOA production sites have been dumped on soil or surface waters. Apart from solvents and acids or bases this waste often contains remainders of the synthesis products, which can then be dissipated in the aquatic environment or seep through the soil into groundwater.

Considering that DOA are highly active in the human body, it can be expected that some of them, in particular the more persistent ones, may exert some effects on aquatic biota when their levels increase in the aquatic environment.

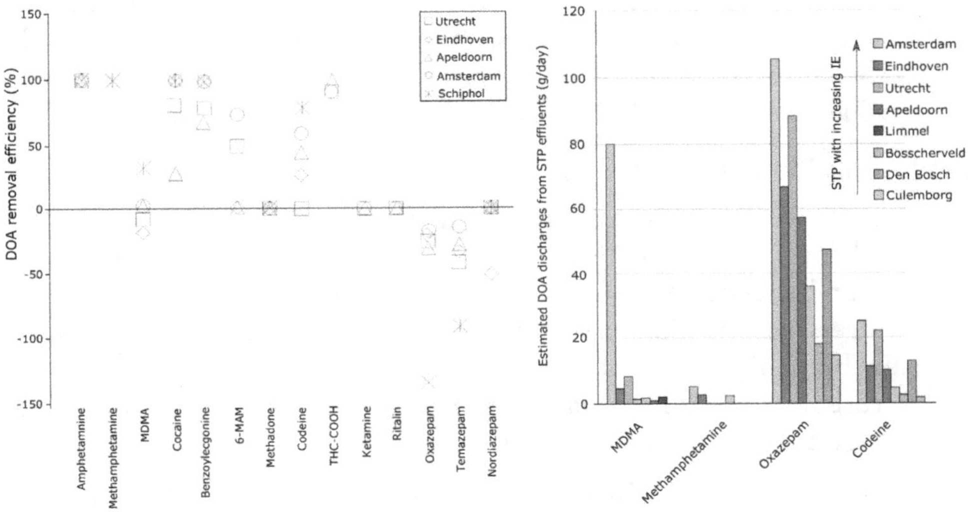

Figure 4. Data demonstrating that WWTP emit DOA to receiving waters. A) Removal efficiencies of DOA recorded in five Dutch WWTPs; B) Estimated discharges (g/day) of DOA from WWTPs based on monitoring data and WWTP effluent flow rates in 2009 (Sources: Bijlsma et al, 2012; Van der Aa et al, 2013). Drawn by Wilma IJzerman.

References

Van der Aa, M., Bijlsma, L., Emke, E., et al. (2013). Risk assessment for drugs of abuse in the Dutch watercycle. Water Research 47(5), 1848-1857.

Bijlsma, L., Emke, E., Hernández, F., de Voogt, P. (2012). Investigation of drugs of abuse and relevant metabolites in Dutch sewage water by liquid chromatography coupled to high resolution mass spectrometry. Chemosphere 89(11), 1399-1406.

EMCDDA, European Drug Report 2018, Lisbon (http://www.emcdda.europa.eu/publications/edr/trends-developments/2018_en)

Thomas, K. V., Bijlsma, L., Castiglioni, S., et al. (2012). Comparing illicit drug use in 19 European cities through sewage analysis. Science of the Total Environment 432, 432-439.

2.3.5. Hydrocarbons

Introduction

Hydrocarbons are a class of chemicals that only consist of carbon and hydrogen atoms. But despite their simplicity in building blocks, this group of chemicals consists of a wide variety of structures, as there are differences in chain length, branching, bonding types and ring structures. The main sources of hydrocarbons are crude oil and coal, which are formed over millions of years by natural decomposition of the remains of plants, animals or wood, and are used to derive products we are using on a daily basis, including fuels and plastics Other natural sources include natural burning (forest fires) and volcanic sources

Hydrocarbon Classification

The major classes of hydrocarbons are paraffins (i.e. alkanes), naphthenics (i.e. cycloalkanes) and aromatics (Figure 1), and within these classes, several subclasses can be identified. Paraffins are hydrocarbons that do not contain any ring structures. Paraffins can be subdivided in normal (n-) paraffins, which do not contain any branching (straight chain), and iso-paraffins (i-), which do contain a branched carbon-chain. When alkanes include at least one carbon-carbon double bond, they are considered olefins (or alkenes).

Naphthenic and aromatic hydrocarbons both contain ring-structures but differ in the presence of aromatic or non-aromatic rings. The naphthenics and aromatics can be further specified based on their ring count; often mono-, di- and poly-ring structures are distinguished from each other. Of all these classes, the polycyclic aromatic hydrocarbons (PAHs) are the best-studied category in terms of all kinds of environmental aspects.

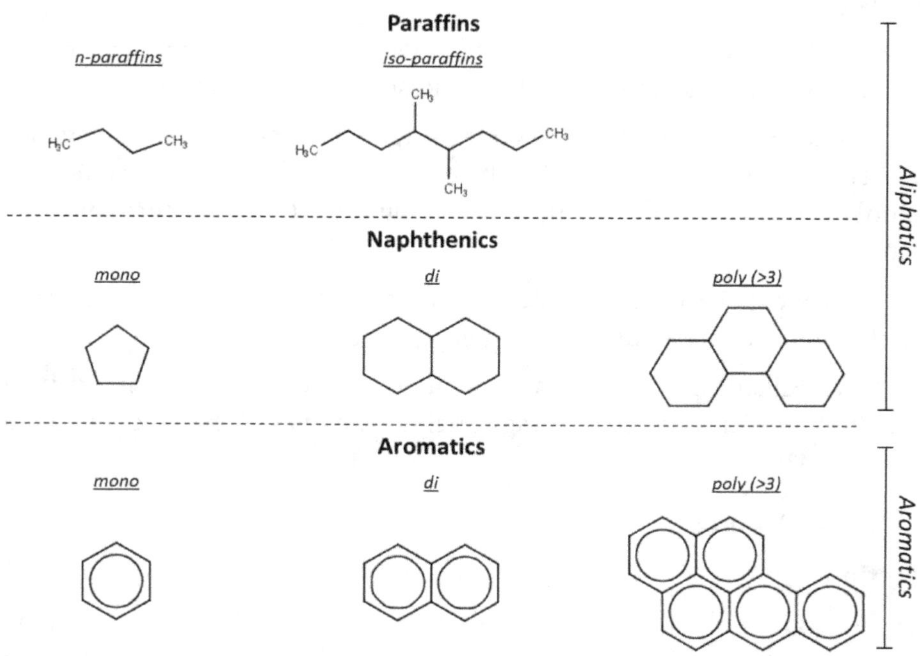

Figure 1. Chemical structures of common hydrocarbon classes. (by author)

Besides the classes considered in Figure 1., combinations of these classes also exist. Naphthenic or aromatic structures with an alkane side chain are mostly still considered as naphthenic or aromatic hydrocarbons, respectively. However, when a non-aromatic-ring is fused with an aromatic-ring, the hydrocarbon is classified as a naphthenic-aromatic structure. Depending on the ring-count several subclasses can be identified, including naphthenic-mono-aromatics and naphthenic-poly-aromatics.

Concerns for Human Health and the Environment

Because of their lack of polar functional groups, hydrocarbons are generally hydrophobic and, as a consequence, many are able to cause acute toxic effects in aquatic animals by a non-specific mode of action known as narcosis (or baseline toxicity). Narcosis is a reversible state of inhibited activity of membrane structures within the cells of organisms. Narcosis type toxicity is considered the minimum toxicity that any substance will be able to have, just by reaching concentration levels in the phospholipid bilayer of the cell membranes that disturb membrane transportation process. Hence the name "baseline" or minimum toxicity.

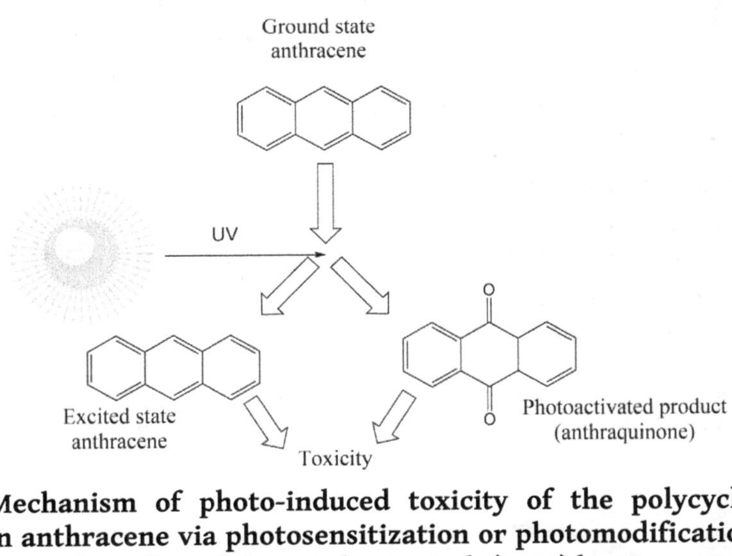

Figure 2. Mechanism of photo-induced toxicity of the polycyclic aromatic hydrocarbon anthracene via photosensitization or photomodification reactions, respectively. Adapted from Roberts et al. (2017) by Steven Droge.

When these events take place above a certain threshold, systemic toxicity can be observed in the organism, such as lethality. This threshold concentration is also known as the critical body residue (CBR) (Bradbury et al., 1989; Parkerton et al., 2000; Veith & Broderius, 1990). Nevertheless, hydrocarbons can also have a more specific mechanisms of action, resulting in greater toxicity than baseline toxicity. For example, the toxicity of several PAHs increases in combination with ultraviolet radiation due to photo-induced toxicity. Photo-induced toxicity may be caused by photoactivation, in which a PAH is degraded into an oxidized product with a higher toxicity, or rather by photosensitization, in which reactive oxygen species (ROS) are formed due to an excited state of the PAHs (Figure 2) (Roberts et al., 2017).

PAHs are especially vulnerable to photodegradation as their absorption spectrum falls within the range of wavelengths reaching the earth's surface (> 290 nm), which is not the case for most monoaromatic and aliphatic hydrocarbons (EMBSI, 2015). The photo-induced effects are of particular concern for aquatic species with transparent bodies, like zooplankton and early life stages, as more UV-light can penetrate into their organs and tissues (Roberts et al., 2017).

Several hydrocarbons are also able to cause genotoxicity and cancer upon exposure, including benzene, 1,3-butadiene and some PAHs. The carcinogenicity of PAHs is caused by biotransformation into reactive metabolites, specifically into epoxides which are the first step in oxidation of aromatic ring structures into dihydrodiol ring systems (Figure 3). In general, the biotransformation step increases the water solubility of the hydrocarbons (Phase I metabolism) and promotes subsequent conjugation and excretion (Phase II metabolism). However, several epoxide metabolites – more specifically the most stable aromatic epoxides - can reach the cell nucleus and covalently react with DNA, forming DNA adducts, and induce mutations (Figure 3). Ultimately, if not repaired such mutations can accumulate and may result in the formation of tumors (Ewa & Danuta, 2016). Specifically, PAHs with a bay-like region are of concern as biotransformation results in relatively stable reactive epoxides that are not accessible to epoxide hydrolase enzymes (Figure 3) (Jerina et al. 1980). Similar to PAHs, 1,3-butadiene and benzene are also able to cause cancer via the effects of their respective reactive metabolites (Kirman et al., 2010; US-EPA 1998).

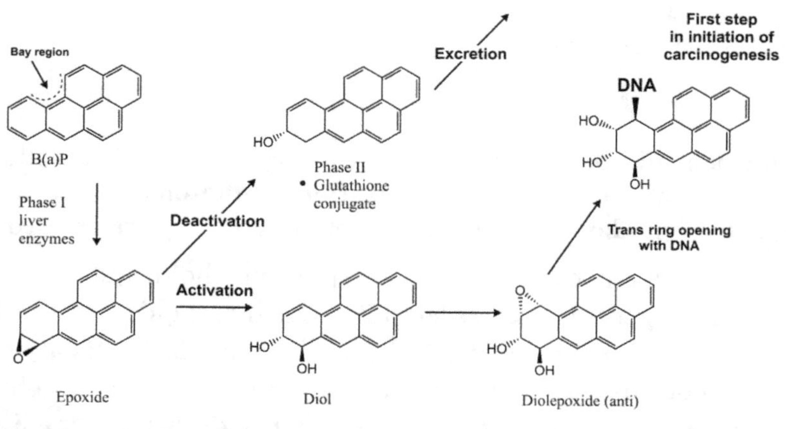

Figure 3. The biotransformation pathways of benzo(a)pyrene and binding to the DNA of reactive intermediates. Adapted from Homburger et al. (1983) by Steven Droge.

Besides their toxicity, some hydrocarbons such as the high molecular weight PAHs can be persistent in the environment and may accumulate in biota as a result of their hydrophobicity. It is therefore expected that internal concentrations are higher for such hydrocarbons and it is interesting that there is thus a relationship between narcosis and bioaccumulation potential. Consequently, these hydrocarbons might be of even greater concern.

Characterization of Mixtures of Hydrocarbons

As most research focused on specific hydrocarbons, including several PAHs, it is important to note that the biodegradation, bioaccumulation and toxicity potential of many hydrocarbons is still not fully known, such as for alkylated PAHs and naphthenics. As there is such a wide variety in hydrocarbon structures, it is impossible to assess the (potential) hazards of all hydrocarbons separately. Therefore, grouping approaches have been developed to speed up the risk assessment. Within a grouping approach, hydrocarbons can be clustered based on structural similarities. The underlying assumption is that all chemicals in a group are expected to have fairly similar physicochemical properties, and subsequently also fairly similar environmental fate and effect properties. As a result, such a group could potentially be assessed as if it is one single hydrocarbon.

#C	Par	iP	MNap	DiNap	PolyNap	MAr	DiAr	PolyAr	NapMAr	NapDiAr	NapPolyAr	...	Total
...	...	...	...	...	...	...	...	...	...	...	...	...	...
11	3%	...	...	...	...	2%	...	...	...	...	...	...	5%
12	8%	4%	2%	1%	...	3%	1%	...	...	...	...	...	19%
13	15%	6%	4%	2%	...	4%	3%	1%	...	...	...	...	35%
14	9%	2%	2%	2%	...	4%	5%	3%	...	...	...	...	27%
15	1%	2%	1%	1%	...	3%	3%	2%	...	...	...	...	13%
16	...	...	1%	...	...	...	...	...	...	...	...	...	1%
17	...	...	...	...	...	...	...	...	...	...	...	...	...
18	...	...	...	...	...	...	...	...	...	...	...	...	...
...	...	...	...	...	...	...	...	...	...	...	...	...	...
Total	36%	14%	10%	6%	...	16%	12%	6%	...	...	...	...	100%

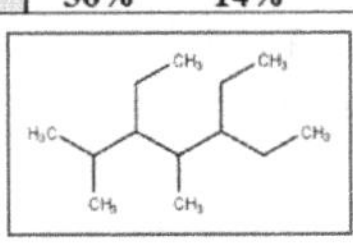

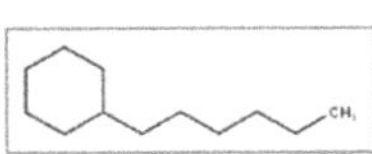

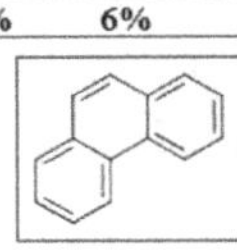

Figure 4. Theoretical example matrix of the hydrocarbon block method based on functional classes (columns) and carbon number (rows). Percentages represents the relative presence of specific hydrocarbon block within an oil substance. The PBT-properties of a block can potentially be assessed by analyzing and extrapolating PBT-properties of representative hydrocarbon structures.

The applicability of a hydrocarbon specific grouping approach, known as the Hydrocarbon Block Method (King et al., 1996), to assess the biodegradation and bioaccumulation potential of hydrocarbons is currently being investigated. Within this approach, all hydrocarbons are grouped based on their functional class (e.g. paraffin, naphthenic, aromatic) and the number of carbon atoms. The number of carbon atoms is thought to highly correlate with the boiling point of the hydrocarbons. An example matrix of the Hydrocarbon Block Method is presented in Figure 4. The composition of an oil substance could be expressed in such a matrix following GC-GC/MS analysis. Subsequently, the PBT-properties of the individual blocks could potentially be assessed by analyzing and extrapolating the PBT-properties of representative hydrocarbons for varying hydrocarbon blocks (see Figure 4).

References

Bradbury, S.P., Carlson, R.W., Henry, T R. (1989). Polar narcosis in aquatic organisms. In Aquatic Toxicology and Hazard Assessment: 12th Volume. ASTM International.

EMBSI (2015). Assessment of Photochemical Processes in Environmental Risk Assessment of PAHs

Ewa, B., Danuta, M.Š. (2017). Polycyclic aromatic hydrocarbons and PAH-related DNA adducts. Journal of applied genetics 58, 321-330.

Homburger, F., Hayes, J.A., Pelikanm E.W. (1983). A Guide to General Toxicology. Karger/Base, New York, NY.

Jerina, D.M., Sayer, J.M., Thakker, D.R., Yagi, H., Levin, W., Wood, A.W., Conney, A.H. (1980). Carcinogenicity of polycyclic aromatic hydrocarbons: the bay-region theory. In Carcinogenesis: Fundamental Mechanisms and Environmental Effects (pp. 1-12). Springer, Dordrecht.

King, D.J., Lyne, R.L., Girling, A., Peterson, D.R., Stephenson, R., Short, D. (1996). Environmental risk assessment of petroleum substances: the hydrocarbon block method. CONCAWE report no. 96/52.

Kirman, C.R., Albertini, R.A., & Gargas, M.L. (2010). 1, 3-Butadiene: III. Assessing carcinogenic modes of action. Critical reviews in toxicology 40(sup1), 74-92.

Parkerton, T.F., Stone, M.A., Letinski, D. J. (2000). Assessing the aquatic toxicity of complex hydrocarbon mixtures using solid phase microextraction. Toxicology letters 112, 273-282.

Roberts, A.P., Alloy, M.M., Oris, J.T. (2017). Review of the photo-induced toxicity of environmental contaminants. Comparative Biochemistry and Physiology Part C: Toxicology & Pharmacology 191, 160-167.

US-EPA (1998). Carcinogenic Effects of Benzene: An Update. EPA/600/P-97/001F.

Veith, G.D., Broderius, S.J. (1990). Rules for distinguishing toxicants that cause type I and type II narcosis syndromes. Environmental Health Perspectives 87, 207.

2.3.6. CFCs

Introduction

CFCs (chlorofluorocarbons) were very common air pollutants in the 20th century because they were the basic components of refrigerants and air conditioning, propellants (in spray can applications), and solvents, since the 1930s. They are still very common air pollutants, because they are very persistent chemicals, and emissions do still continue. In the first years as refrigerants, they replaced the much more toxic components ammonia (NH3), chloromethane (CH3Cl), and sulfur dioxide (SO2). Particularly the CFCs leaking from old refrigerating systems in landfills and waste disposal sites caused high emissions into the environment. Typically, these volatile CFC chemicals are based on the smallest carbon molecules methane (CH4), ethane (C2H6), or propane (C3H8). All hydrogen atoms in these CFC molecules are replaced by a mixture of chlorine and fluorine atoms.

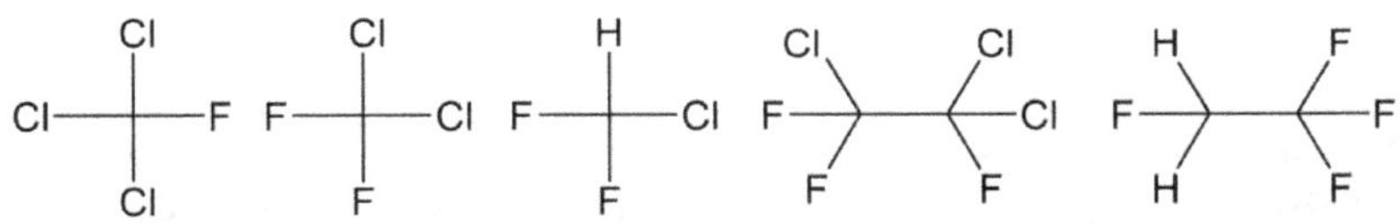

freon-11 BP 23.8°C | freon-12 BP -29.8°C | freon-22 BP -40.8°C | freon-113 BP 47.5°C | freon-134 BP -26.1°C

Figure 1. Different common refrigerants and their boiling points. Freon 134 is chlorine free. . (Source: Steven Droge)

CFCs are less volatile than their hydrocarbon analogue, because the halogen atoms polarize the molecules, which causes stronger intermolecular attractions. Depending on the substitution with Cl or F, the boiling point can be tuned to the desired point for refrigerating cooling processes. The CFCs are also much less flammable than hydrocarbon analogues, making them much safer in all kinds of applications.

Naming of CFCs

CFCs were often known by the popular brand name Freon. Freon-12 (or R^{-12}) for example stands for dichlorodifluoromethane (CCl_2F_2, boiling point -29.8 °C, while methane has -161 °C), as shown in Figure 1. The naming reflects the amount of fluor atoms as the most right number. The next value to the left is the number of hydrogen atoms plus 1, and the next value to the left is the number of carbon atoms less one (zeroes are not stated), and the remaining atoms are chlorine. Accordingly, Freon-113 could apply to 1,1,2-trichloro-1,2,2-trifluoroethane (C2Cl3F3, boiling point 47.7 °C, while ethane has -161 °C). The structure of any Freon-X number can also be derived from adding +90 to the value of X, so Freon-113 would give a value of 203. The first numerical is the number of C (2), the second numerical H (0), the third numerical F (3), and the remaining substitutions are by chlorine (C_2X_6 gives 3 chlorines).

The Reason CFC Depletes the Ozone Layer

The key issue with CFC emissions is the reaction under influence of light ("photodegradation") that ultimately reduces ozone concentrations ("ozone depletion") in the upper atmosphere ("stratosphere"). Ozone absorbs the high energy radiation of the solar UV-B spectrum (280–315nm), and the ozone layer therefore prevents this to reach the Earth's surface. The even more energetic solar UV-C spectrum (100-280nm) is actually causing the formation of ozone (O_3) when reacting with oxygen (O_2), as shown in Figure 2. Under the influence of intense light-energy in the upper atmosphere, CFC molecules can disintegrate into two highly reactive radicals (molecules with a free electron .), for Freon-11:

$$CCl_3F \xrightarrow{h\nu} CCl_2F\cdot + Cl\cdot$$

It is the radical Cl. that catalyzes the conversion of ozone back into O2. The environmentally relevant role of the fluorine atoms in CFCs is that they make these chemicals very persistent after emission, because the C-F bond is one of the strongest covalent bonds known. With half-lives up to >100 years, high CFC levels can reach the upper atmosphere. James Lovelock was the first to detect the widespread presence of CFCs in air in the 1960s, while the damage caused by CFCs was discovered only in 1974. Another undesirable effect of CFC in the stratosphere is that they are a much more potent greenhouse gases than CO_2.

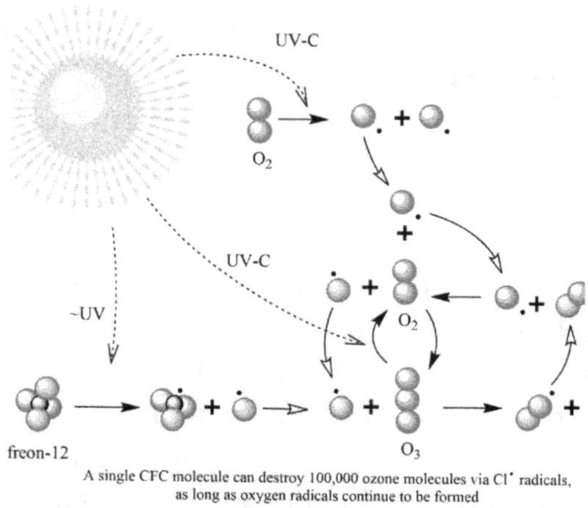

Figure 2. The influence of UV on formation of chlorine radicals from CFCs, oxygen radicals from O2 and oxygen radicals from disintegration of O3. Ozone is not formed in the absence of UV (night time), but can still be reacting away by chlorine radicals. (Source: Steven Droge)

CFC Replacements

In 1978 the United States banned the use of CFCs such as Freon in aerosol cans. After several years of observations of the ozone layer depletions globally (Figure 3), particularly above Antarctica, the Montreal Protocol was signed in 1987 to drastically reduce CFC emissions worldwide. CFCs were banned by the late 1990s in most EU countries, and e.g. in South Korea by 2010. Due to the persistency of CFCs it may take until 2050-2070 before the ozone layer will return to 1980 levels (which were bad already).

The key damaging feature of CFCs in terms of ozone depletion is their persistency, so that emissions reach and build up in the stratosphere (starting from 20km above the equator, but only at 7km above the poles). CFC replacement molecules were initially found simply by adding more hydrogens in the CFC structures and somewhat less Cl (HCFCs), but fractions still contributed to Cl. radicals. Later alternatives lack the chlorine atoms and have even shorter lifetimes in the lower atmosphere, and simply cannot form the Cl radicals. These "hydrofluorocarbons" (HFCs) are currently common in automobile air conditioners, such as Freon-134 (do the math to see that there is no Cl, boiling point -26.1 °C).

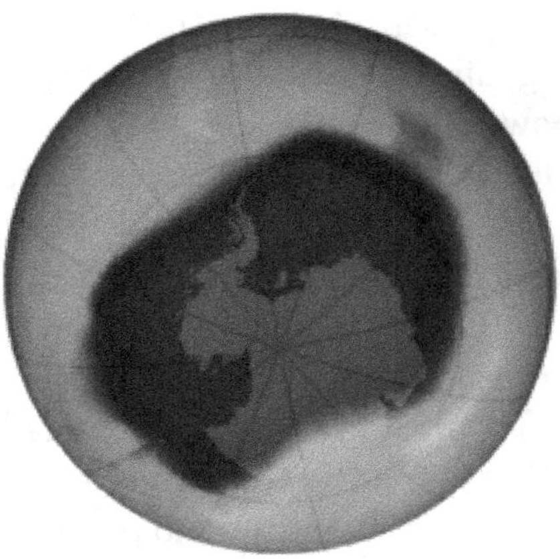

Figure 3. The 2006 record size hole in the ozone layer above Antarctica (Source https://en.wikipedia.org/wiki/Ozone_depletion)

Still, HCFC as well as HFCs are still very potent greenhouse gasses, so the worldwide use of such chemicals remains problematic and gives rise to new legislations, regulations, and searches for alternatives. R-410A (which contains only fluorine) is becoming more widely used but is 1700 times more potent than CO_2 as greenhouse gas, equal to Freon-22. Simple hydrocarbon mixtures such as propane/ isobutane are already used extensively in mobile air conditioning systems, as they have the right thermodynamic properties for some uses and are relatively safe. Unfortunately, we did not have the technological skills, nor the awareness to apply this back in the 1930s.

2.3.7. Cosmetics/Personal Care Products

Introduction

Personal care products (PCPs) cover a large range of products fulfilling hygiene, health, or beauty purposes (e.g. shampoo, toothpaste, nail polish). They are categorized into oral care, skin care, sun care, hair care, decorative cosmetics, body care and perfumes. Overall, most PCPs are classified as cosmetics and regulated accordingly. In the European Union (EU) the Cosmetic Regulation governs the production, safety of ingredients and the labelling and marketing of cosmetic products. The United States of America (USA), on the other hand, have a narrower definition of cosmetics so that products not fulfilling the definition are regulated as pharmaceuticals (e.g. sunscreen) (Food and Drug Administration, 2016).

PCPs come in a range of formats (e.g. liquids, bars, aerosols, powders) and typically contain a wide range of chemicals, each fulfilling a specific function within the product. For example, a shampoo can include cleansing agents (surfactants), chemicals to ensure product stability (e.g. preservatives, pH adjusters, viscosity controlling agents), diluent (e.g. water), perfuming chemicals (fragrances), and chemicals to influence the product's appearance (e.g. colourants, pearlescers, opacifiers). The chemicals present in PCPs ultimately enter the environment either through air during direct use, such as the propellants in aerosols, or through wastewater via down the drain disposal following product use (e.g. shower products, toothpaste). The release of PCP chemicals into the environment needs to be monitored and the safety of these chemicals understood in order to avoid potential problems. In developed countries, the use of wastewater treatment plants (WWTPs) is key to effectively removing the PCP chemicals and other pollutants from wastewater prior to their release to rivers and other watercourses. The removal mechanisms occurring in WWTPs include biodegradation, sorption onto solids, and volatilization to the air. The extent of removal is influenced by the physicochemical properties of the chemicals and the operational conditions of the WWTPs. In regions where wastewater treatment is lacking, the chemicals in PCPs enter the environment directly.

The wide scale daily use of PCPs and the associated large volumes of chemicals released explain why they are scrutinized by environmental protection agencies and regulatory bodies. The following sections will briefly review some of the classes of chemicals used in PCPs by describing their behavior in the environment and their potential effect on ecosystems.

Cleansing Agents - Surfactants

Surfactants are an important and widely used class of chemicals. They are the key components of many household cleaning agents as well as PCPs, such as shampoos, soaps, bodywash and toothpaste, because of their ability to remove dirt. These dirt-removing properties also make surfactants inherently toxic to aquatic organisms. The biodegradability of surfactants is a key legal requirement for their use in PCPs to minimize the likelihood of unsafe levels in the environment. Different types of surfactants exist and are often classified based on their surface charge. Anionic surfactants, which carry a negative surface charge, interact and help remove positively charged dirt particles from surfaces such as hair and skin. Sodium lauryl sulfate is a typical example of an anionic surfactant used in PCPs.

Cationic surfactants, such as cetrimonium chloride, are positively charged and may be used as hair conditioning agents to make hair shinier or more manageable. Non-ionic surfactants (uncharged), such as cetyl alcohol, help formulate products or increase foaming. Amphoteric surfactants, such as sodium lauriaminodipropionate, carry both positive and negative charges and are commonly used to counterbalance the potentially irritating properties of anionic surfactants.

Fragrances

Fragrances are mixtures of often more than 20 perfumery chemicals used to provide the smell of PCPs. Typically, fragrances are present at very low levels in most PCPs (below 0.01%) so that their exact compositions are not disclosed. Disclosed, however, are any allergens present in the fragrance to help dermatologists and consumers avoid certain fragrance chemicals. Despite the wish to protect trade secrets, a recent trend increasingly sees companies disclose the full fragrance compositions of their products on their websites (e.g. L'Oréal, Unilever). Well-known examples of fragrances include hexyl cinnamal, linalool, and limonene. Potential concerns about the ecotoxicological impact of fragrances have arisen on the one hand because of a lack of disclosure of fragrance formulations and on the other hand because of the detection of certain persistent fragrances in the environment (e.g. nitromusks).

Preservatives

Preservatives are usually added to PCPs containing water for their ability to protect the product from contamination by bacteria, yeasts, and molds during storage or repeated use. Given their targeted action against living organisms, the use of preservative in chemical products including PCPs is under constant scrutiny. For example, in 2016 and 2017, the European Commission tightened the regulation around the use of methylisothiazolinone in cosmetics products due to human safety concerns. Other preservatives that have been restricted in use, because of both human safety and environmental safety concerns (e.g. endocrine disruption effects), include certain types of parabens and triclosan.

UV Filters

UV filters are used in sunscreen products as well as in other PCPs such as foundation, lipstick, or moisturizing cream to protect users from UV radiation. UV filters can be organic or inorganic. Inorganic UV filters, like titanium oxide and zinc oxide, form a physical boundary protecting the skin from UV radiation. Organic UV filters, on the other hand, protect the skin by undergoing a chemical reaction with the incoming UV radiation. Organic UV filters commonly found in PCPs include butyl methoxydibenzoylmethane, ethylhexyl methoxycinnamate, and octocrylene. Organic UV filters are poorly biodegradable and have the potential to accumulate in organisms. Further, a number of organic UV filters have been shown to be toxic to coral organisms in laboratory tests. They are suspected to cause coral bleaching by, for example, promoting viral infections but research is still on-going to understand their potential ecotoxicological effects at realistic environmental concentrations.

Volatile Chemicals

Certain chemicals used in PCPs are highly volatile and may end up in the air following product use. Examples include propellants, such as propane butane mixes or compressed air/nitrogen, used in aerosols to apply ingredients in hairsprays or deodorants and antiperspirants. Fragrances also volatilize when the product is applied to skin or hair to provide smell. Volatile silicones, chemicals used to assist the deposition of ingredients in liquids and creams, are another example of chemicals emitted to air upon PCP use.

The Special Case of Plastic Microbeads

Plastic microbeads, with a diameter smaller than 5mm, have been used in PCPs such as face scrubs or shower gels for their scrubbing and cleansing properties. The growing concern about plastic pollution in water has drawn attention to the use of microbeads in PCPs. As a result, a number of initiatives were launched both to highlight the use of plastic microbeads and to encourage replacement with natural alternatives. An example thereof is the "Beat the microbead" coalition (https://www.beatthemicrobead.org/) sponsored by the United Nations Environment Program, launched to help consumers identify and avoid PCPs containing microbeads. Such initiatives together with voluntary commitments by industry have led to a large decrease in the use of microbeads in wash-off cosmetic products: In the EU, for example, the use of microbeads in wash-off products was reduced by 97% from 2012 to 2017. Legislation to restrict the use of microbeads has also recently been put in place. In the USA microbeads in PCPs were banned in July 2017 and a number of EU countries (e.g. United Kingdom, Italy) have also banned their use in wash-off products.

Further Reading

For more information on PCP chemicals and their function in products, please see (European commission 2009; Grocery Manufacturers Association 2017).

For more information on the different types of surfactants, please see Tolls et al. (2009) and Section 2.3.8.

Manova et al. (2013) list the different types of UV filters.

The report of Scudo et al. (2017) gives more information on the use of microplastics in Europe.

References

European Commission (2019). Cosing. 2009 03.2019]; Available from: http://ec.europa.eu/growth/tools-databases/cosing/.

Food and Drug Administration (2016). Are All "Personal Care Products" Regulated as Cosmetics? [cited 2019 03]; Available from: https://www.fda.gov/forindustry/fdabasicsforindustry/ucm238796.htm.

Grocery Manufacturers Association (2017). Smartlabel. [cited 2017 11]; Available from: http://www.smartlabel.org/.

Manova, E., von Goetz, N., Hauri, U., Bogdal, C., Hungerbuhler, K. (2013). Organic UV filters in Personal Care Products in Switzerland: A Survey of Occurrence and Concentrations. International Journal of Hygiene and Environmental Health 216, 508-514.

Scudo, A., Liebmann, B., Corden, C., Tyrer, D., Kreissig, J., Warwick, O. (2017). Intentionally Added Microplastics in Products. in: Limited A.F.W.E.a.I.U., ed. United Kingdom

Tolls, J., Berger, H., Klenk, A., Meyberg, M., Müller, R., Rettinger, K., Steber, J. (2009). Environmental safety aspects of Personal Care Products - a European perspective. Environmental Toxicology and Chemistry 28, 2485-2489.

2.3.8. Detergents and Surfactants

Introduction

Surface active agents ("surf-act-ants") are a wide variety of chemicals produced in bulk volumes (>10.000 tonnes annually) as a key ingredient in cleaning products: detergents. Typical for surfactants is that they have a hydrophobic tail and a hydrophilic head group (Figure 1). At relatively high concentrations in water (typically >10-100 mg/L), surfactants spontaneously form aggregated structures called micelles (Figure 1), often in spheres with the hydrophobic tails inward and the hydrophilic head groups towards the surrounding water molecules.

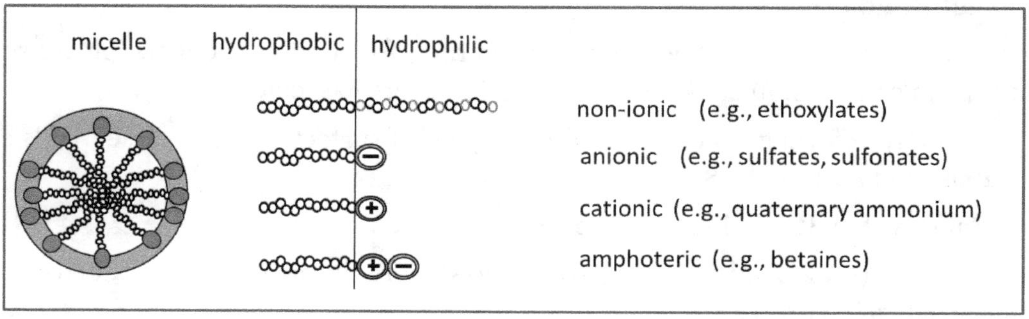

Figure 1. Different forms (micelle and surfactant monomer) and types of surfactants. (Source: Steven Droge)

These micelle super-structures allow surfactants to dissolve grease and dirt from e.g. textile or dishes into water, which can then be flushed away. Besides this common use of surfactants, their amphiphilic (i.e., both hydrophilic and lipophilic) properties allow for a versatile use in our modern world:

- During the large 2010 oil spill in the Mexican Gulf, enormous volumes (>6700 tonnes) of several types of surfactant formulations (e.g. "Corexit") were used to disperse the constant stream of oil leaking from the damaged deep water well into small dissolved droplets, in order to facilitate microbial degradation and prevent the formation of floating oil slabs that could ruin coastal habitats.
- The ability of a layer of surfactants to maintain hydrophobic particles in solution is a key process in many products, such as paints and lacquers.
- The ability to emulsify dirt particles is a key feature in process fluids during deep drilling in soil or sediment.
- Fabric softners, and hair conditioners, have cationic surfactants as key ingredients that stick with the positively charged head groups onto the negatively charged fibers of your towel or hair. After the final flushing, these cationic surfactants still stick on the fibers and because of the hydrophobic head groups sticking out make these materials feel soft and smooth. Often only during the next washing event (with anionic or nonionic surfactants) the cationic surfactants are flushed off the fibers.
- Many cationic surfactants have biocidal properties at relatively low concentrations and are therefore used in a few percent in many cosmetic products as preservatives, e.g. in cosmetics, or used to kill microbes in food processing, antibacterial hand wipes or during swimming pool cleaning. Examples are chloride salts of benzalkonium, benzethonium, cetylpyridinium.

- Surfactants lower the surface tension of water, and therefore are used (as "adjuvants") in pesticide products to facilitate the droplet formation during spraying and to improve contact of the droplets with the target leaves in case of herbicides. Examples are fluorinated surfactants, silicone based surfactants (Czajka et al. 2015), and polyethoxylated tallow amine (POEA) used for example in the glyphosate formulation Roundup.

The hydrophobic tail of surfactants is mostly composed of a chain of carbon atoms, although also fluorinated carbon ($-CF_2-$) chains or siloxane ($Si(CH_3)_3$-O-[..]-$Si(CH_3)_3$) chains are also possible.

The first bulk volume produced surfactants for washing machines were branched anionic alkylbenzenesulfonates (ABS) and alkylphenolethoxylates (APEO), with the hydrocarbon source obtained from petroleum. Because of the variable petroleum source, these chemicals are often complex mixtures. However, hydrophobic branched alkylchains are poorly biodegraded, and the constant disposal of these surfactants into the waste water caused very high environmental concentrations, often leading to foaming rivers (Figure 2).

Figure 2. Foaming in sewage treatment and on rivers in the 1950s caused by non-biodegradable tetrapropylene sulfonate (TPS) (Source: Kümmerer 2007, who obtained permission from the Archives of the Henkel Company, Düsseldorf).

Surfactant producers 'voluntarily' switched to carbon sources such as palm oil or controlled polymerization of petrol-based ethylene, that could be used to generate surfactants with linear alkyl chains: linear alkylbenzenesulfonate (LAS) and alcohol ethoxylates (AEO). Some surfactants have the hydrophilic headgroup attached to two carbon chains, such as the anionic docusate (heavily used in the BP oil spill) and the cationic dialkyldimethylammonium chemicals. Common detergent surfactants are nowadays designed to pass ready biodegradability tests (>60% mineralisation to CO_2 within a 10 d window following a lag phase, in a 28 d test). Early examples of fabric softners are double chain (dialkyl)dimethylammonium surfactants, but the environmental persistency of these compounds (DODMAC and DHTDMAC, see e.g. EU and RIVM-reports) has led to a large replacement by diesterquats (DEEDMAC), which degrade more rapidly through the weak ester linkages of the fatty acid chains (Giolando et al. 1995). A switch to sustainable production of the carbon sources is ongoing. Whereas petroleum based ethylene oil was mostly used, it is being replaced increasingly by the linear fatty acid carbon chains from either palm-oil (mostly C_{16}/C_{18}), coconu oil (mostly C_{12}/C_{14}), but also such raw materials needs to be as sustainably derived as possible.

The hydrophilic headgroups can vary extensively. Nonionic surfactants can have a simple polar functional group (amide), glucose based (polyglycoside), or contain a variable lengths of repetitive ethoxlyate and/or propoxylate units. Because the ethoxylation process is difficult to control, such surfactants are often complex mixtures. Anionic surfactants are often based on sulfate (SO_{4-}) or sulfonate (SO_4-), but also phosphonate and carboxylates are common. A key difference between anionic surfactants is that sulfate and sulfonates are fully anionic (pK_a ~<0) over the entire environmental pH range (pH4-9), while carboxylates are weaker acids that are still partially neutral species (pK_a ~5). Most cationic surfactants are based on permanently charged quaternary ammonium headgroups ($R\text{-}(N+)(CH_3)_3$), although several ionizable amine groups are applied in cationic surfactants too (e.g., diethanolamines).

The key ingredient property of most surfactants is the critical micelle concentration (CMC), which defines the dissolved concentration above which micellar aggregates start to form that can remove grease or fully emulsify particles. The CMC decreases proportionally with the hydrophobic tail length, and this means that with longer tails, you need less surfactant to start to form micelles. However, with increasing hydrophobic tails anionic surfactants more readily precipitate with dissolved inorganic cations such as calcium. Also, surfactant toxicity increases proportionally with hydrophobic tail lengths. If the alkyl chain is too long, the surfactant may bind strongly to all kinds of surfaces and not be available for micelle formation. The optimum hydrophobic chain length is thus often a balance between the desired properties of the surfactant and several critical processes that influence the efficiency and risk of surfactants.

References

Kümmerer, K. (2007). Sustainable from the very beginning: rational design of molecules by life cycle engineering as an important approach for green pharmacy and green chemistry. Green Chemistry 9, 899–907 . DOI: 10.1039/b618298b

Giolando, S.T., Rapaport, R.A. , Larson, R.J., Federle, T.W., Stalmans, M., Masscheleyn P. (1995). Environmental fate and effects of DEEDMAC: A new rapidly biodegradable cationic surfactant for use in fabric softeners. Chemosphere 30, 1067-1083. DOI: 10.1016/0045-6535(95)00005-S

Czajka, A., Hazell, G., Eastoe, J. (2015). Surfactants at the Design Limit. Langmuir 31, 8205–8217. DOI: 10.1021/acs.langmuir.5b00336

Scientific Committee on Toxicity, Toxicity and the Environment (CSTEE) (2001). OpinionontheresultsoftheRiskAssessmentof:Dimethyldioctadecylammonium chloride (DODMAC). EU-report C2/JCD/csteeop/DodmacHH22022002/D(02) https://ec.europa.eu/health/archive/ph_risk/committees/sct/documents/out143_en.pdf

Van Herwijnen, R. (2009). Environmental risk limits for DODMAC and DHTDMACRIVM Letter report 601782029 - https://www.rivm.nl/bibliotheek/rapporten/601782029.pdf

Chapter 3 ENVIRONMENTAL CHEMISTRY, FROM FATE TO EXPOSURE

If you want to re-use this chapter, for e.g. in your electronic learning environment, feel free to copy this url: maken.wikiwijs.nl/120175/3__Environmental_Chemistry___From_Fate_to_Exposure

3.1. ENVIRONMENTAL COMPARTMENTS

3.1.1. Introduction

In order to understand and predict the effects of chemicals in the environment we need to understand the behaviour of chemicals in specific environments and in the environment as a whole. In order to deal with the diversity of natural systems, we consider them to consist of compartments. These are defined as parts of the physical environment that are defined by a spatial boundary that distinguishes them from the rest of the world, for example the atmosphere, soil, surface water and even biota. These examples suggest that three phases: gas, liquid, and solid, are important but compartments may consist of different phases. For example, the atmosphere consists of suspended liquids (e.g., fog) and solids (e.g., dust) as well as gases. Similarly, lakes contain suspended solids and soils contain gaseous and water-filled pore space. In detailed environmental models, each of these phases may also be considered to be a compartment.

The behaviour and fate of chemicals in the environment is determined by the properties of environmental compartments and the physicochemical characteristics of the chemicals. Together these properties determine how chemicals undergo chemical and biological reactions, such as hydrolysis, photolysis and biodegradation, and phase transfer processes such as air-water exchange and sorption.

In this chapter, we first introduce the most important compartments and their most important properties and processes that determine the behaviour of chemical contaminants: the atmosphere, the hydrosphere, sediment, soil, groundwater and biota. The emissions of chemicals into the environment from either point sources or diffuse sources is discussed and the important pathways and processes determining the fate of chemicals. The partitioning approach to phase-transfer processes is presented with sorption as a specific example. The impact of physicochemical properties on partitioning is also discussed.

Other important environmental processes are discussed in sections on metal speciation, processes affecting the bioavailability of metals and organic contaminants and the transformation and degradation of organic chemicals. These sections also include information on the basic methods to measure these processes.

Finally, approaches that are used to model and predict the environmental fate of chemicals, and thus the exposure of organisms to these chemicals are described in section 3.8.

3.1.2. Atmosphere

Composition and Vertical Structure of Atmosphere

The atmosphere of the Earth consists of several layers that have limited interaction. The troposphere is the lowermost part of the atmosphere and contains the oxygen that we breathe and most of the water vapor. It contains on average 78% N_2, 20% O_2 and up to 4% water vapor. Greenhouse gases like CO_2 and CH_4 are present at 0.0038 % and 0.0002%, respectively. Air pollutants like ozone and NO_2 have concentrations that are even a factor 1,000-10,000 lower, but are already harmful for the health of humans, animals and vegetation at these concentrations.

The troposphere is 6-8 km high near the poles, about 10 km at mid-latitudes and about 15 km at the equator. It has its own circulation and determines what we experience as weather, with temperature, wind, clouds and precipitation. The lowest part of the troposphere is the boundary layer, the part that is closest to the Earth.

Its height is determined by the heating of the atmosphere by the Earth surface and the wind conditions and has a daily cycle, determined by the incoming sunlight. It is not a completely separate layer, but the exchange of air pollutants like O_3, NOx, SO_2, and xenobiotic chemicals between the boundary layer and the above layers is generally inefficient. Therefore it is also termed the mixing layer.

Above the troposphere there is the stratosphere, a layer that is less strongly influenced by the daily solar cycle. It is very dry and has its owns circulation, with some exchange with the troposphere. The stratosphere contains the ozone layer that protects life on Earth against UV radiation and extends to about 50 km altitude. The layers covering the next 50 km are the mesosphere and thermosphere, which are not directly relevant for the transport of the chemicals considered in this book.

Properties of Pollutants in The Air

Air pollutants include a wide range of chemicals, ranging from metals like lead and mercury to asbestos fibers, polycyclic hydrocarbons (PAH) and chloroform. These pollutants may be emitted into the atmosphere as a gas or as a particle or droplet with sizes of a few nanometer to tens of micrometers. The particles and droplets are termed aerosol, or, depending on the measurement method, particulate matter. The latter definition is used in air quality regulations. Note that a single aerosol can be composed of several chemical compounds. Once a pollutant is released in the atmosphere, it is transported by diffusion and advection by horizontal and vertical winds and may be ultimately deposited to the Earth's surface by rain (wet deposition), and by sticking to the surface (dry deposition). Large particles may fall down by gravitational settling, a process also called sedimentation. Air pollutants may interact with each other or with other chemicals, particles and water by physical or chemical processes. All these processes will be explained in more detail below. A summary of the relevant interactions is given in Figure 1.

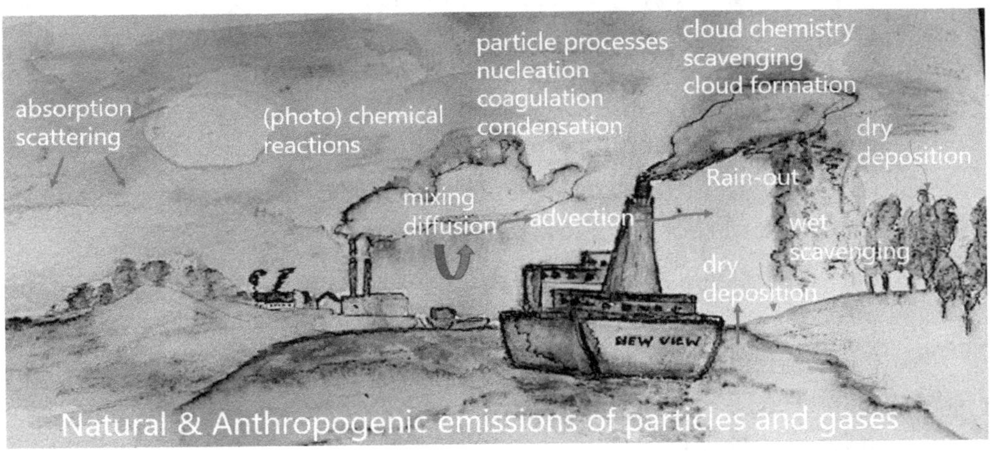

Figure 1. Overview of the most relevant process in the atmosphere related to release and transport of air pollutants (source: Wilma IJzerman).

It is important to realize that air pollutants can have an impact on meteorology itself, by acting as a greenhouse gas, scattering or absorbing incoming light when in aerosol form, or be involved in the formation of clouds. This aspect will not be discussed here.

Meteorology is relevant for all aspects, ranging from mixing and transport to temperature or light dependent reaction rates and absorption of water. Depending on the removal rates, species may be removed with timescales of seconds, like heavy sand particles, to decades or longer, like halogen (Cl, Br)-containing gases, and be transported over ranges of a few meters to crossing the globe several times. Concentrations of gases are often expressed in volume mixing ratios (parts per billion, ppb) whereas for particulate matter the correct unit is in (micro)gram per cubic meter as there is no molecular weight associated to it. For ultrafine particles, concentrations are expressed as numbers of particles per cubic meter, for asbestos, the number of fibers per cubic meter is used.

Physical and Chemical Processes Determining the Properties of Air Pollutants

The properties of air pollutants, like solubility in water, attachment efficiencies to the Earth's surface (water, vegetation, soil) and size of particles, are key elements determining the lifetime and transport distances. These properties may change due to the interaction with other chemicals and with meteorology.

The main physical processes are:

- Condensation or evaporation with decreasing/increasing temperature. A potentially toxic aerosol may thus be covered by semi-volatile chemicals like ammonium sulfate, whilea gas may condensate on an aerosol and be transported further as part of this aerosol. Some pollutants exist at the same time in aerosol and in gas phase with their partitioning depending on air temperature and relative humidity.
- Gases may cluster to form ultrafine particles of a few nanometers (nucleation) that will grow to larger sizes.
- Particles may grow by coagulation: rapidly moving small particles bump into large slow-moving particles and remain attached to it.
- Particles may take up water (hygroscopicity), leading to a larger diameter.

Chemical conversions include:

- Chemical reactions between gas-phase pollutants and ambient gas, which alters the characteristics of an air pollutant or can lead to the formation of pollutants (e.g. NO_2 being directly emitted by combustion, leading to ozone formation).
- Chemical reactions between aerosols and gases, often involving water attached to aerosol.
- Cloud droplets or water attached to aerosols have their own role in the chemistry of the atmosphere, and gases may diffuse into the water .
- Some pollutants may act as a catalyst.
- Some air pollutants may be degraded by (UV) light (photodegradation).

Pollutants are characterized by their chemical composition but for aerosols also the size distribution of particles is relevant. Note that the conservation of atoms always applies, but particle size distribution and particle number can be changed by physical processes. This has to be kept in mind when concentrations are expressed in particles per volume instead of mass concentrations.

Transport of Air Pollutants

Several processes determine the mixing and transport of chemicals in the air:

- Diffusion due to the motion of molecules, or Brownian motion of particles (random walk).
- Turbulent diffusion: the mixing due to (small-scale) turbulent eddies which have a random nature, but the strength of this diffusion is related to friction in the flow.
- Advection: the process of transport with the large-scale flow (wind speed and direction).
- Mixing or entrainment of different air masses leads to further mixing of air pollutants over a larger volume. This process is for example relevant when the sun rises, and air in the boundary layer is heated, rises, and mixes with the air in the layer above the boundary layer.

Although the processes of diffusion and transport are well-known, it is not an easy task to solve the equations describing these processes. For stationary point and line sources under idealized conditions, analytical descriptions can be derived in terms of a plume with concentration profile with a Gaussian distribution, but for more realistic descriptions the equations must be solved numerically. For complex flow around a building, computational fluid dynamics is required for an accurate description, for long-range transport a chemistry-transport model must be used.

Wet Deposition

Wet deposition comprises the removal processes that involve water:

- Scavenging of particles by or dissolution of gases in cloud droplets. These cloud droplets may grow to larger size and rain out, thereby removing the dissolved air pollutants from the atmosphere (in-cloud scavenging).
- Particles below the clouds may be scavenged by falling raindrops (below-cloud scavenging).
- Occult deposition occurs when clouds are in contact with the surface (mountain areas) and cloud droplets containing air pollutants stick to the surface.

Wet deposition is a very efficient removal mechanism for both small (<0.1 μm diameter) and large aerosols (diameter >1 μm). Aerosols that are hygroscopic can grow in size by absorbing water, or shrink by evaporating water under dry conditions. This affects their deposition rate for wet or dry deposition.

Dry Deposition

Dry deposition is partly determined by the gravitational forces on a particle. Heavy particles (≥ 5 μm) fall to the Earth's surface in a process called gravitational settling or sedimentation. In the lowest layer of the atmosphere, air pollutants can be brought close enough to the surface to stick to it or be taken up. In the turbulent boundary layer, air pollutants are brought close to the surface by the turbulent motion of the atmosphere, up to the very thin laminar layer (laminar resistance, only for gases) through which they diffuse to the surface. Aerosols or gases can stick to the Earth's surface or be taken up by vegetation, but they may also rebound. Several pathways take place in parallel or in series, similar to an electric circuit with several resistances in parallel and series. Therefore the resistance approach is often used to describe these processes.

Deposition above snow or ice is generally slow, since the atmosphere above it is often stably stratified with little turbulence (high *aerodynamic resistance*), the surface area to deposit on is relatively small (impactors) and aerosols may even rebound to an icy surface (collection efficiency of impactors) to which it is difficult to attach. On the other hand, forests often show high deposition velocities since they induce stronger turbulence in the lowermost atmosphere and have a large leaf surface that may take up gases by the stomata or provide sticking surfaces for aerosols.

Deposition velocities thus depend on the type of surface, but also on the season, atmospheric stability (wind speed, cloud coverage) and ability of stomata to take up gases. When the atmosphere is very dry, for example, plants close their stomata and this pathway is temporarily shut down. For particles, the dry deposition velocity is lowest at sizes of 0.1-1 μm.

Re-emission

Once air pollutants are removed from the atmosphere, they can be part of the soil or water compartments which can act as a reservoir. This is in general only taken into account for a limited number of chemicals. Ammonia or persistent organic pollutants may be re-emitted from the soil by evaporation. Dusty material or pollutants attached to dust may be brought back into the atmosphere by the action of wind. This is relevant for bare areas like agricultural lands in wintertime, but also for passing vehicles that bring up the dust on a road by the flow they induce.

Atmospheric Fate Modelling

Due to the many relevant processes and interactions, the fate of chemical pollutants in the air has to be determined by using models that cover the most important processes. Which processes need to be covered depends on the case study: a good description of a plume of toxic material during an accident, where high concentrations, strong gradients and short timescales are important, requires a different approach than the chronic small release of a factory. Since it would require too heavy numerical simulations to include all aspects, one has to select the relevant processes to be included. Key input for all transport models are emission rates and meteorological input.

When one is interested in concentrations close to a specific source, next to emission rate the effective emission height is important, and processes that determine dispersion: wind speed, atmospheric stability. Chemical reaction rates and deposition velocities should be included when the time horizon is long or when the reactions are fast or deposition velocities are high.

When one is interested in actual concentrations resulting from releases of multiple sources and species over a large area of interest, like for an air quality forecast, the processes of advection, deposition and chemical conversions become more relevant, and input meteorology needs to be known over the area. Sharp gradients close to the individual sources are, however, no longer resolved. In particular rain can be a very efficient removal mechanism, removing most of the aerosol within one hour. Dry deposition is slower, but results in a lifetime of less than a week and transport distances of less than 1,000 km for most aerosols. For some gaseous compounds like halogens and N_2O deposition does hardly play a role and they are chemically inert in the troposphere, leading to very long lifetimes.

To assess the overall long-term fate of a new chemical to be released to the market, the potential concentrations in air, water and soil have to be determined. Ideally, models for air, soil and water are used together in a consistent way, including their interaction For many air pollutants the atmospheric lifetime is short but determines where and in which form they are deposited onto ground and water surfaces, where they may accumulate. This means that even if a concentration in air is relatively low at a certain distance from a source, the deposition of an air pollutant over a year may still be significant. Figure 2 shows an example of annual mean modelled concentrations and annual total deposition of a hypothetical passive (non-reactive) soot-like tracer that is released at 1 kg/hour at a fictitious site in The Netherlands. Annual mean concentrations are small compared to ambient concentrations of particulate matter, but the footprint of the accumulated deposition is larger than that of the mean concentration, since the surface acts as a reservoir. This implies that re-emission to air can be relevant. It may take several years for soil or water before an equilibrium concentration is reached in these compartments from the deposition input, as different processes and time scales apply. Mountain ranges are visible in the accumulated wet deposition (Alps, Pyrenees), as they are areas with enhanced precipitations.

In addition to spatially explicit modelling, also box models exist that have the advantage that they can make long-term calculations for a continuous release of a species, including interaction between the compartments air, soil and water. They can be used to determine when an equilibrium concentration is reached within a compartment, but these models cannot resolve horizontal concentration gradients within a compartment.

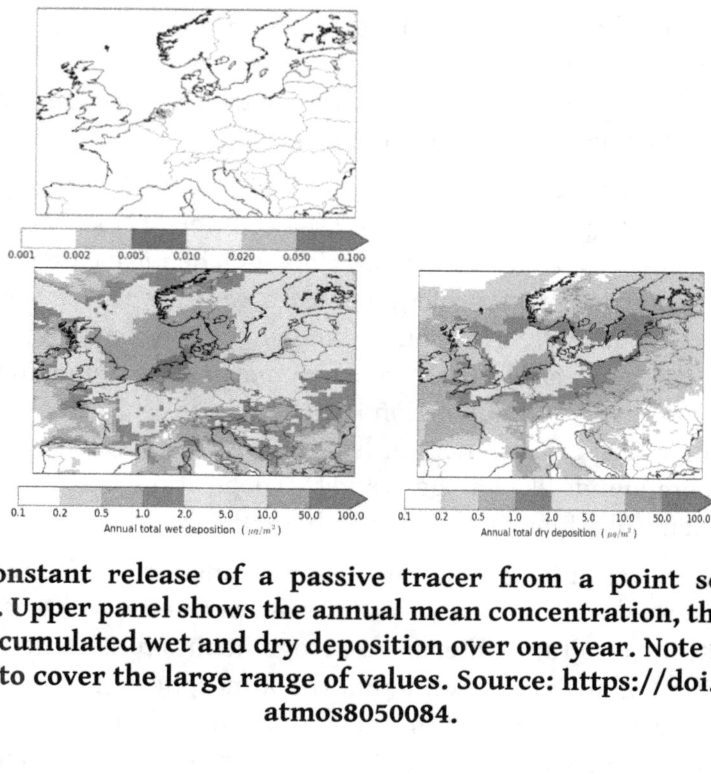

Figure 2. Constant release of a passive tracer from a point source in The Netherlands. Upper panel shows the annual mean concentration, the lower panel shows the accumulated wet and dry deposition over one year. Note the nonlinear colour scale to cover the large range of values. Source: https://doi.org/10.3390/atmos8050084.

Learn More

EU air quality, policy and air quality legislation: http://ec.europa.eu/environment/air/index_en.htm

US hazardous air pollutants, including lists of toxics: https://www.epa.gov/haps

Plume dispersion approach: http://courses.washington.edu/cewa567/Plumes.PDF

Chemistry-transport models: https://www.narsto.org/sites/narsto-dev.ornl.gov/files/Ch71.3MB.pdF

Seinfeld, J., Pandis, S.N. Atmospheric Chemistry and Physics, from air pollution to climate change, Wiley, 2016, covering all aspects.

John, A. C., Küpper, M., Manders-Groot, A. M., Debray, B., Lacome, J. M., Kuhlbusch, T. A. (2017). Emissions and possible environmental implication of engineered nanomaterials (ENMs) in the atmosphere. Atmosphere, 8(5), 84.

3.1.3. Hydrosphere

The Properties and Importance of Water

Water covers 71% of the earth's surface and this water, together with the smaller amounts present as gas in the atmosphere, as groundwater and as ice is referred to collectively as the hydrosphere. The bulk of this water is salt water in the oceans and seas with only a minor part of freshwater being present as lakes and rivers (Figure 1).

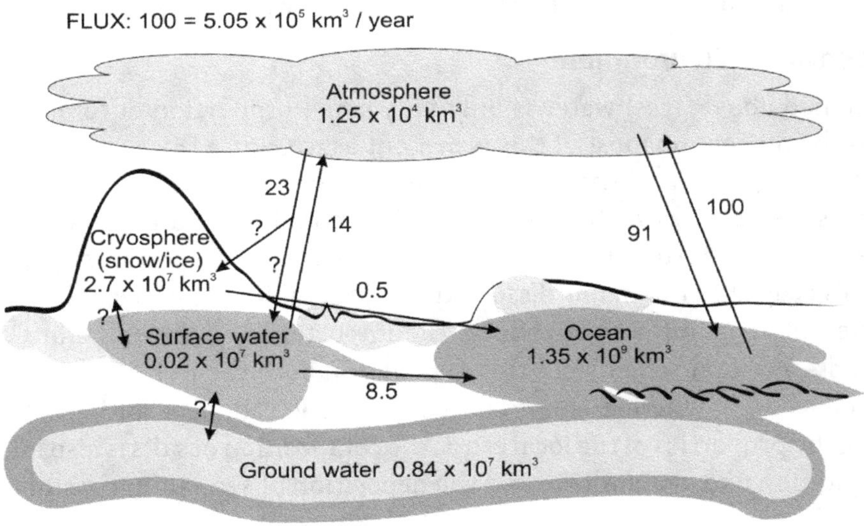

Figure 1. Global hydrological cycle and water balance (arrows are fluxes of water per year). Adapted from Kayane (1992) and Peixoto (1994) by Steven Droge 2019.

Water is essential for life and also plays a key role in many other chemical and physical processes, such as the weathering of minerals and soil formation and in regulating the Earth's climate. These important roles of water derive from its structure as a small but very polar molecule arising from the polarised hydrogen-oxygen bonds (Figure 2). As a consequence, water molecules are strongly attracted by hydrogen bonding, giving it relatively high melting and boiling points, heat capacity, surface tension, etc. The polarity of the water molecule also makes water an excellent solvent for a wide variety of ionic and polar chemicals but a poor solvent for large nonpolar molecules.

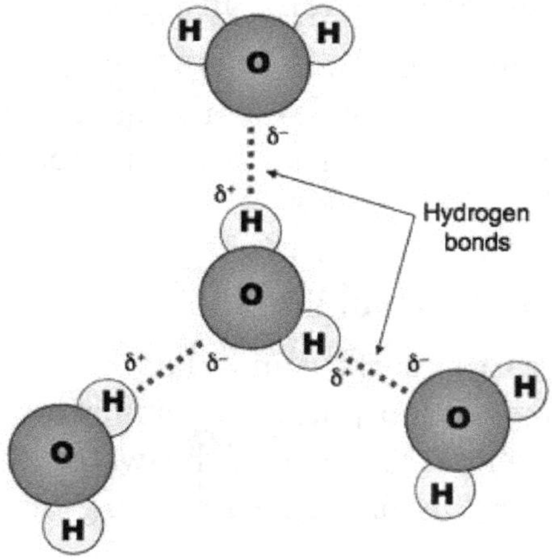

Figure 2. Hydrogen bonding between water molecules

The Freshwater Environment

As mentioned above, freshwater is only very small proportion of total amount of water on the planet and most of this is present as ice. Since this water is in contact with the atmosphere and the soils and bedrock of the Earth's crust, it dissolves both atmospheric gases such as oxygen and carbon dioxide and salts and organic chemicals from the crust. If we compare the relative compositions of cations in the Earth's crust and the major dissolved species (Table 1) it is clear that these are very different. This difference reflects the importance of the solubility of these components. For ionic chemicals, this depends on both their charge and their size (expressed as z/r2, where z is the charge and r the radius of an ion). As well as reflecting the properties of the local crust, the composition of salts is also influenced by precipitation and evaporation and the deposition of sea salt in coastal regions.

Table 1. Comparison of the major cation composition of average upper continental crust and average river water. (*except aluminum and iron from Broecker and Peng (1982))

	Upper continental crust (mg/kg) (Wedepohl 1995*)	River water (mg/kg) (Berner & Berner 1987*)
Al	77.4	0.05
Fe	30.9	0.04
Ca	29.4	13.4
Na	25.7	5.2
K	28.6	1.3
Mg	13.5	3.4

The pH of surface water is determined by both the dissolution of carbonate minerals and carbon dioxide from the atmosphere. These components are part of the set of equilibrium reactions known as the carbonate system (Figure 3).

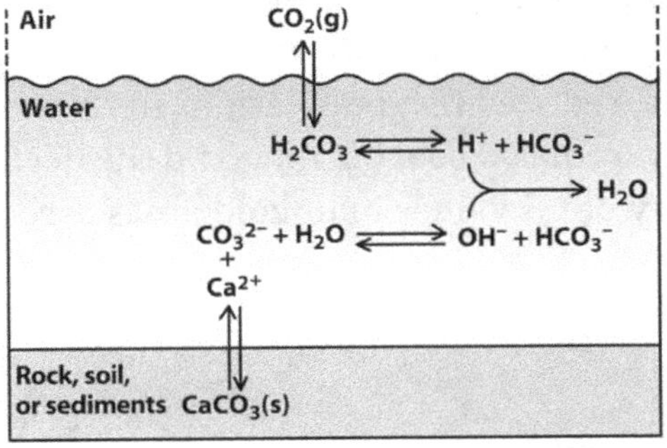

Figure 3. The carbonate system of equilibria regulating the pH of surface water. Source: http://butane.chem.uiuc.edu/pshapley/GenChem1/L26/3.html

At equilibrium with the current atmospheric CO_2 concentration and solid calcium carbonate, the pH of surface water is between 7 and 9 but this may reach more acidic values where soils are calcium carbonate (limestone) poor. This is illustrated by the pH values measured in a river in Northern England, where acidic, organic carbon-rich water at the source is gradually neutralised once the river encounters limestone rich bedrock (Figure 4).

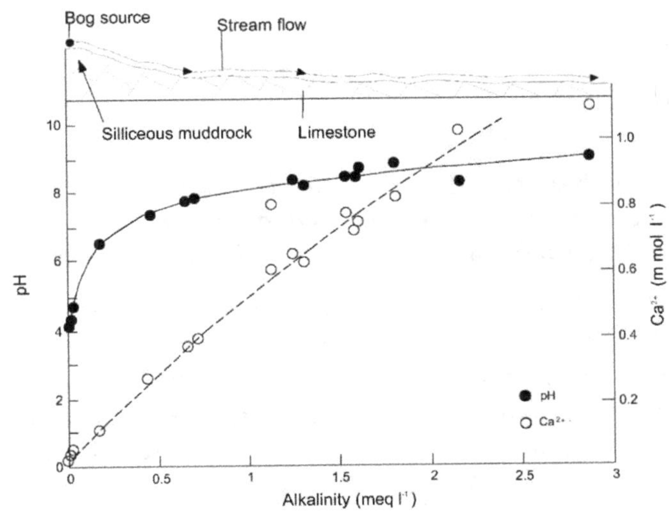

Figure 4. Water chemistry in the Malham Tarn area of northern England, showing the relationship between pH, alkalinity and dissolved calcium as water flows from bog on siliceous mudrock to limestone where the pH is buffered around 8 due to once limestone weathering. Redrawn from Fig. 5.6 in Andrews et al. (2004) by Wilma Ijzerman.

As well as these natural processes, there are human influences on the pH of surface water including acidic precipitation resulting from fossil fuel combustion and acidic effluents from mining activities caused by oxidation and dissolution of mineral sulphides. Regions such as Southern Scandinavia with carbonate-poor soils are particularly vulnerable to acidification due to these influences and this is reflected in for example, reduced fish populations in these vulnerable regions (see Figure 5). More recently, reduced coal burning and the decline in heavy industry is resulting in the recovery of pH values in upland areas across Europe.

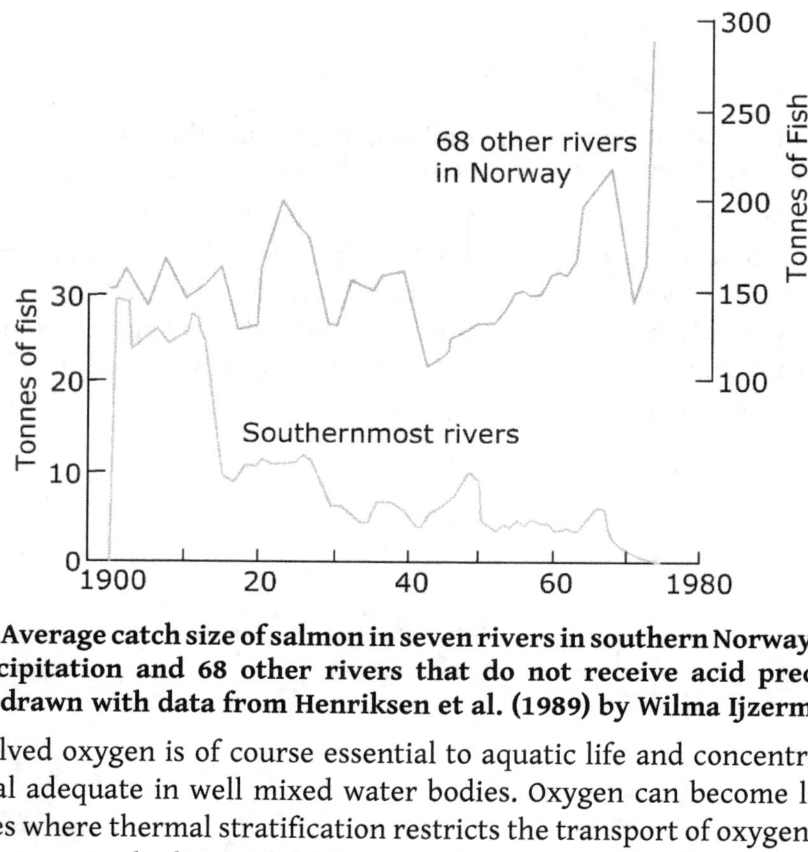

Figure 5. Average catch size of salmon in seven rivers in southern Norway receiving acid precipitation and 68 other rivers that do not receive acid precipitation. Redrawn with data from Henriksen et al. (1989) by Wilma Ijzerman.

Dissolved oxygen is of course essential to aquatic life and concentrations are in general adequate in well mixed water bodies. Oxygen can become limiting in deep lakes where thermal stratification restricts the transport of oxygen to deeper layers, or in water bodies with high rates of organic matter decomposition. This may result in anoxic conditions with significant ecological impacts and on the behaviour of chemical contaminants.

The Marine Environment

Freshwater eventually moves into seas and oceans where the concentrations of dissolved species are much higher than in the freshwater environment. This is partly due to the effects of evaporation of water from the oceans but is also be due to specific marine sources of some dissolved components. Estuaries are the transition zones where freshwater and seawater mix. These are highly productive environments where increasing salinity has a major impact on the behaviour of many chemicals, for example on the speciation of metals and the aggregation of colloids as a result of cations shielding the negative surface change of colloidal particles (Figure 6). Increasing salinity also affects organic chemicals, with ionic chemicals forming ion pairs, and even reducing the solubility of neutral organics (the so-called salting-out effect). As well as these chemical effects due to increasing salinity, the lowering of flow rates in estuaries leads to the deposition of suspended particles.

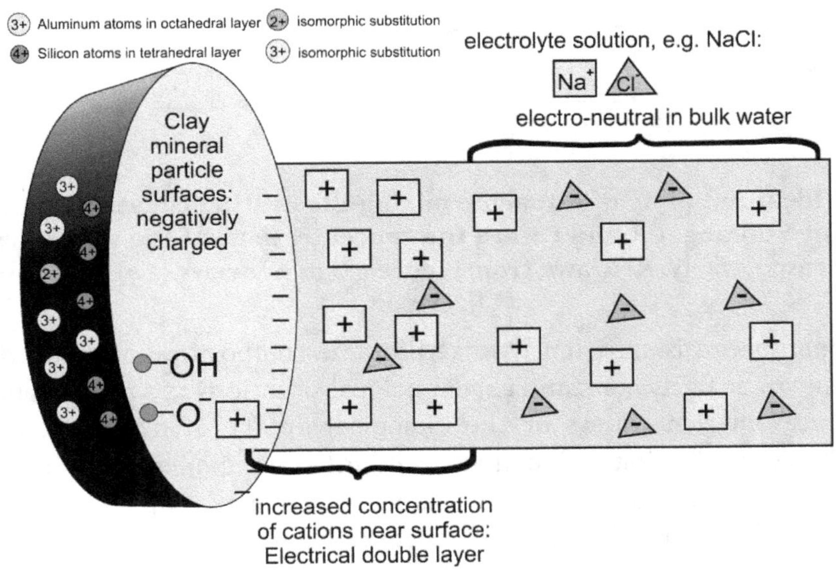

Figure 6. The Electrical Double Layer (EDL), comprising a fixed layer of negative charge on a clay particle (due to isomorphic substitutions and surface acids) and a mobile ionic layer in solution. The latter is caused because positive ions are attracted to the particle surface. Note that with increasing distance from the particle surface the solution approaches electrical neutrality. (Source Steven Droge 2019)

Since the concentrations of pollutants are in general lower in the marine environment than in the freshwater environment, concentrations in estuaries decrease as freshwater is diluted with seawater. Measuring salinity at different locations in estuaries is a convenient way to determine the extent of this dilution. Components that are present in higher concentrations in seawater will of course show an increase with salinity unless. Plotting salinity against the concentrations of chemicals at different locations can yield information on whether they behave conservatively (i.e. only undergoing mixing) or are removed by processes such as degradation or partitioning into the atmosphere or sediments. Figure 7 shows examples of plots expected for conservative chemicals and those that are either removed in the estuary or have local sources there. Models describing the behaviour of chemicals in estuaries can be used with these data to derive the rates of removal or addition of the chemical in the system.

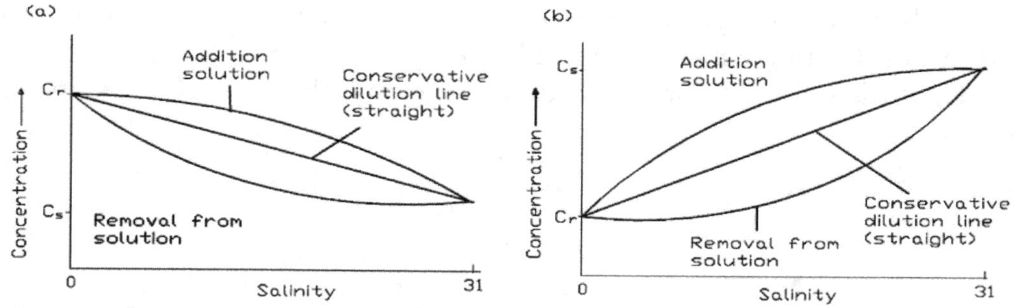

Figure 7. Idealized plots of estuarine mixing illustrating conservative and non-conservative mixing. CR and CS are the concentrations of the ions in river and seawater respectively. Redrawn from Figure 6.3 in Andrews et al. (2004) by Wilma Ijzerman.

The open ocean is sufficiently mixed for the composition of major dissolved constituents to be fairly constant, except in local situations as a result of upwelling of deep nutrient-rich waters or the biological uptake of nutrients. In coastal regions the concentrations of chemicals and other components originating from terrestrial sources may also be locally higher. The major components in seawater are listed in Table 2 with their typical concentrations.

Table 2. Major ion composition of freshwater and seawater.

	Seawater (mmol/L) (Broecker and Peng, 1982)	River water (mmol/L) (Berner and Berner, 1987)
Na+	470	0.23
Mg2+	53	0.14
K+	10	0.03
Ca2+	10	0.33
HCO3-	2	0.85
SO42-	28	0.09
Cl-	550	0.16
Si	0.1	0.16

These concentrations may be higher in waterbodies that are partly or wholly isolated from the oceans and are impacted by evaporative losses of water (e.g. Mediterranean, Baltic, Black Sea). In extreme case, concentrations of salts may exceed their solubility product, resulting in precipitation of salts in evaporate deposits.

As is the case in freshwater, carbonates play an important role in regulating the ocean pH. The fact that the oceans are supersaturated in calcium carbonate makes it possible for a variety of organisms to have calcium carbonate shells and other structures. The important processes and equilibria involved are illustrated in Figure 8. There is concern that one of the most important effects of increasing atmospheric carbon dioxide will be lowering of ocean pH to values that will result in destabilisation of these carbonate structures.

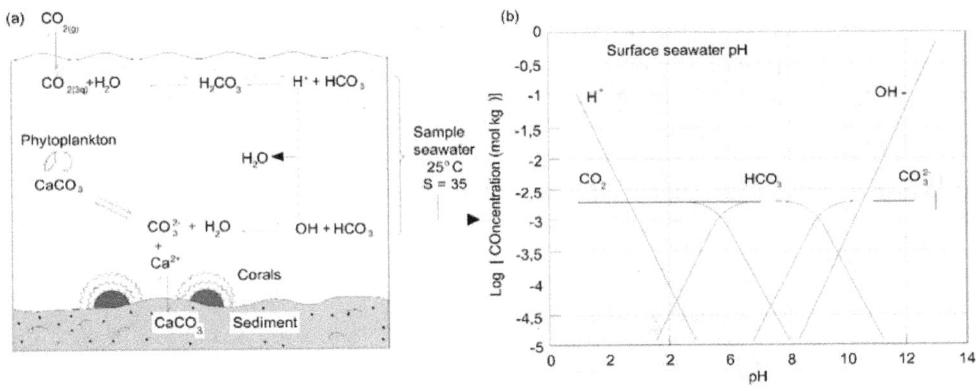

Figure 8. (a) Schematic diagram to illustrate the buffering effect of $CaCO_3$ particles (suspended in the water column) and bottom sediments on surface water HCO_3^- concentrations (after Baird and Cann 2012). (b) A sample of the seawater in (a) will have a pH very close to 8 because of the relative proportions of CO_2, HCO_3^- and CO_3^{2-}, which in seawater is dominated by the HCO_3^- species. Increased CO_2 concentrations in the atmosphere from anthropogenic sources could induce greater dissolution of $CaCO_3$ sediments including coral reefs. Redrawn from Figure 6.8 in Andrews et al. (2004) by Wilma Ijzerman.

References

Andrews, J.E., Brimblecombe, P., Jicketts, T.D., Liss, P.S., Reid, B.J. (2004). An Introduction To Environmental Chemistry, Blackwell Publishers, ISBN 0-632-05905-2.

Baird, C., Cann, M. (2012). Environmental Chemistry, Fifth Edition, W.H. Freeman and Company, ISBN 978-1429277044.

Berner, E.K., Berner, R.A. (1987). Global water cycle: geochemistry and environment, Prentice-Hall.

Broecker, W.S., Peng, T.S. (1982). Tracers in the Sea, Lamont-Doherty Geol. Obs. Publ.

Henriksen, A., Lien, L., Rosseland, B.O., Traaen, T.S., Sevaldrud, I.S. (1989). Lake Acidification in Norway: Present and Predicted Fish Status. Ambio 18, 314-321

Wedepohl, K.H. (1995). The composition of the continental crust, Geochimica Cosmochimica Acta 59, 1217-1232.

3.1.4. Sediment

In preparation

3.1.5. Soil

Introduction

Soil is the upper layer of the terrestrial environment that serves as a habitat for organisms and medium for plant growth. In addition, it also plays an important role in water storage and purification and helps to regulate the Earth's atmosphere (e.g. carbon storage, gas fluxes, ...).

Soils are composed of three phases (Figure 1).

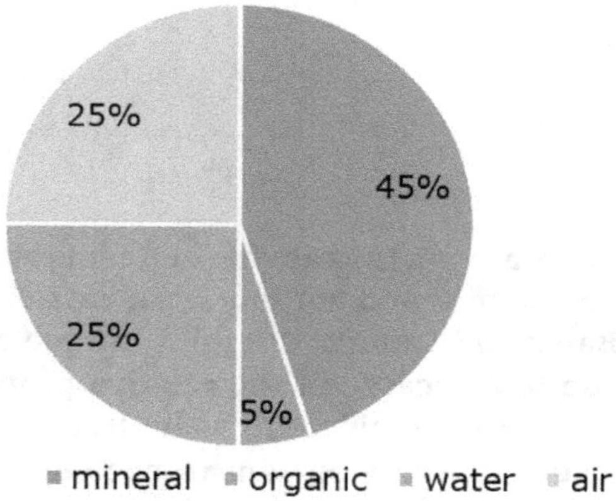

Figure 1. Average composition of soil (in volume %).

The *solid phase* is formed by mineral and organic components. Mineral components appear in different particle sizes from coarse particles (sand), intermediate (silt) and fine (clay) which combination determine soil texture. The particles can be arranged to form porous aggregates; soil pores being filled with air and/or water. The proportion of air in soils depends on soil moisture content. The composition of the soil solid phase may be quite variable.

The gaseous phase has a similar composition as the air, but due to the respiration of plant roots and the metabolic activity of soil microorganisms, O2 content generally is lower and CO2 content higher. Exchange of gases between soil pores and atmospheric air takes place by diffusion. Diffusion proceeds faster in dry soil and much slower when soil pores are filled with water.

The liquid phase of the soil, the soil solution or pore water, is an aqueous solution containing ions (mainly Na^+, K^+, Ca^{2+}, Cl^-, NO_3^-, SO_4^{2-}, HCO_3^-) from dissolution of a variety of salts, and also contains dissolved organic carbon (DOC, also referred to as dissolved organic matter, DOM). The soil solution is part of the hydrological cycle, which involves input from among others rain and irrigation, and output by water uptake by plants, evaporation, and drainage to ground and surface water. The soil solution acts as a carrier for the transport of chemicals in soil, both to plant roots, soil microorganisms and soil animals and to ground and surface water.

Soil Solids

The soil solid phase consists of mineral and organic soil particles. Based on their size, the mineral particles are divided into sand (63-2000 μm), silt (2- 63 μm), and clay (<2 μm). With increasing particle size, the specific surface area decreases, pore size increases and water retention capacity decreases. The sand fraction mainly consists of quartz (SiO_2) and does not have any sorption properties because the quartz crystals are electrically neutral. Sandy soils have large pores, so a low capacity to retain water. In soils with a high silt fraction, smaller pores are better represented, giving these soils a higher water retention capacity. Also the silt fraction has no adsorptive properties. Clays are aluminium silicates, lattices composed of SiO_4 tetrahedrons and $Al(OH)_6$ octahedrons. Upon the formation of clay particles, isomorphic substitution occurred, a process in which Si_4^+ was replaced by Al^{3+}, and Al^{3+} by Mg^{2+}. Although having similar diameters, these elements have different valences. As a consequence, clay particles have a negative charge, making positive ions to accumulate on their surface. This includes ions important for plant growth, like NH^{4+}, K^+, Na^+ and Mg , but also cationic metals (Figure 2). Many other minerals have pH-dependent charges (either positive or negative) which are also important in binding cations and anions.

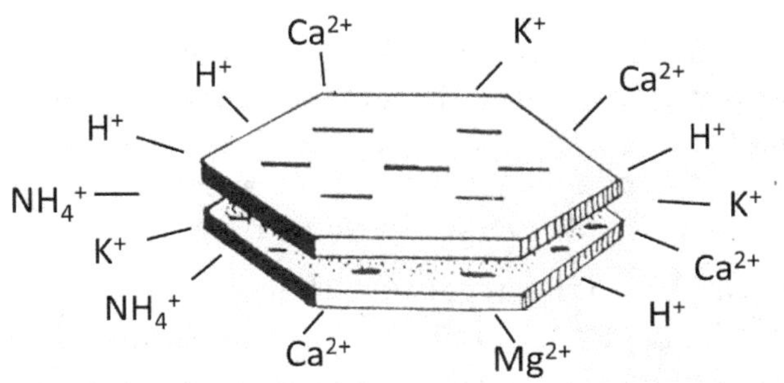

Figure 2. Schematic representation of a clay particle. Due to its negative charge, cations will accumulate near the surface of the clay particle.

In addition to mineral particles, soils also contain organic matter, which includes all dead plant and animals remains and their degradation products. Living biota is not included in the soil organic matter faction. Organic matter is often divided into: 1. humin, non-dissolved organic matter associated with clay and silt particles, 2. humic acids having a high degree of polymerization, and 3. fulvic acids containing more phenolic and carboxylic acid groups. Humic and fulvic acids are water soluble but their solubility depends on pH. For example, humic acids are soluble at alkaline pH but not at acidic pH. The dissociation of the phenolic and carboxylic groups gives the organic matter also a negative charge (Figure 3), the density of which increases with increasing soil pH. The soil organic matter acts as a reservoir of nitrogen and other elements, provides adsorption site for cations and organic chemicals, and supports the building of soil aggregates and the development of soil structure.

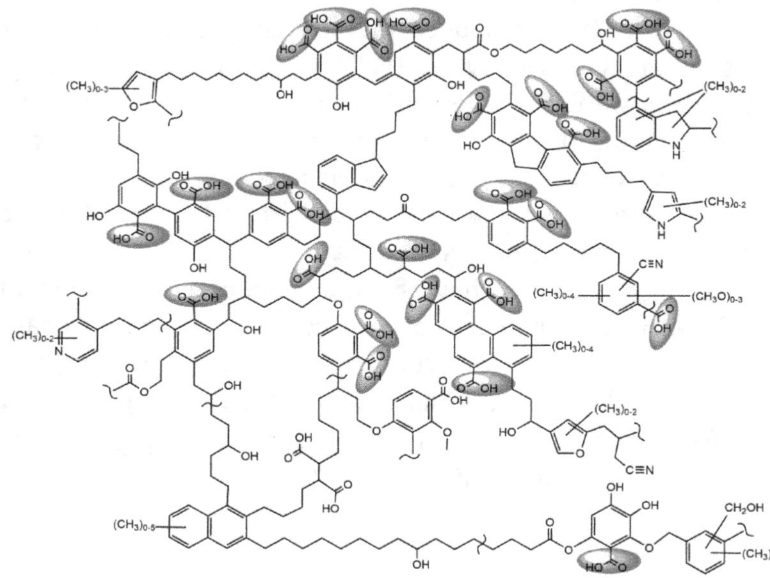

Figure 3. Proposed structure of a humic acid molecule. The phenolic and carboxylic acid groups on the molecule may dissociate depending on the pH of the soil, giving rise to negative sites on the molecule. With this, humic acid contributes to the Cation Exchange Capacity of the soil. Adapted from Schulten & Schnitzer (1997) by Steven Droge.

The binding of cations to the negatively charged sites on the soil particles is an exchange process. The degree of cation accumulation near soil particles depends on their charge density, the affinity of the cations to the charged surfaces (which is higher for bivalent than for monovalent cations), the concentration of ions in solution (the higher the concentration of a cation in solution, the higher attraction to soil particles), etc. Due to their binding to charged soil particles, cations are less available for leaching and for uptake by organisms. The Cation Exchange Capacity (CEC) is commonly used as a measure of the number of sites available for the sorption of cations. CEC is usually expressed as cmolc/kg dry soil. Soils with higher CEC have a higher capacity to bind cations, so cationic metals show a lower (bio) availability in high CEC soils (see the Section on metal speciation). CEC depends on the content and type of clay minerals, with montmorillonite having a higher CEC than e.g. kaolinite, organic matter content and pH of the soil. In addition to clay and organic matter, also aluminium and iron oxides and hydroxides may contribute to the binding of cations to the soil.

Soil Water

The transport of water through soil pores is controlled by gravity, and by suction gradients which are the result of water retention by capillary and osmotic processes. Capillary binding of water is stronger in smaller soil pores, which explains why clayey soils have higher water retention capacities than sandy soils. The osmotic binding of water increases with increasing ionic strength, and is especially high close to charged soil particles like clay and organic matter where ions tend to accumulate.

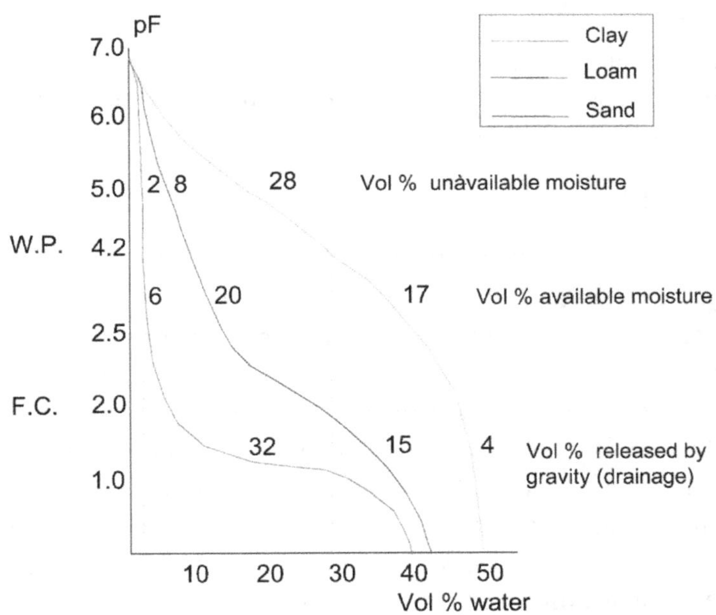

Figure 4. pF curves showing the retention of water by three different soil types. pF is the log of the force with which water is retained, expressed in hPa. W.P. = wilting point, F.C. = field capacity. Source: Wilma IJzerman.

The stronger water is retained by soil, the lower its availability is for plants and other organisms. The strength by which water is retained depends on moisture content, because 1. at decreasing moisture content the ionic strength of the soil solution and therefore osmotic binding increases, 2. when soil moisture content decreases the larger soil pores will be emptied first, leading to increasing capillary retention of the remaining water in smaller pores. Water retention curves describe the strength with which water is retained as a function of total water content and in dependence of the composition of the soil. Figure 4 shows pF curves for three different soil types.

A pF value of 2.2-2.5 corresponds with a binding strength of 200 to 300 hPa. This is called field capacity; water is readily available for plant uptake. At pF 4.2 (15,000 hPa), water is strongly bound in the soil and no longer available for plant uptake; this is called the wilting point. For soil organisms, not the total water content of a soil is of importance but rather the content of available water. Water retention curves may be important to describe the availability of water in soil. Toxicity tests with soil organisms are typically performed at 40-60% of the water holding capacity (WHC) of the soil, which corresponds with field capacity.

References/Further Reading

Schulten, H.-R., Schnitzer, M. (1997). Chemical model structure for soil organic matter and soils. Soil Science 162, 115-130.

Blume, H.-P., Brümmer, G.W., Fleige, H., Horn, R., Kandeler, E., Kögel-Knabner, I., Kretzschmar, R., Stahr, K., Wilke, B.-M. (2016). Scheffer/Schachtschabel Soil Science, Springer, ISBN 978-3-642-30941-0

3.1.6. Groundwater

Introduction

Some definitions conceive all water beneath the earth's surface as groundwater while others restrict the definition to water in the saturated zone. In the saturated zone the pores are completely filled with water in contrast to the undersaturated zone in which some pores are filled with gas and capillary action are important for moving water. Geological formations, which host groundwater in the saturated zone, can be classified as 'aquifer', ' aquitard', or 'aquifuge' depending on their permeability. In contrast to aquitard and aquifuge, which have a low permeability, an aquifer permits water to move in significant rates under ordinary field conditions. Aquifers typically have a high porosity and the pores are well connected with each other. Examples of aquifers include sedimentary layers of sand or gravel, carbonate rocks, sandstones, volcanic rocks and fractured igneous rocks. The redox chemistry discussed in this chapter is focusing on aquifers in sedimentary formations.

Groundwaters are an important source for drinking water and the quality of groundwater is, therefore, of high importance for protecting human health.

However, aquifers also represent a habitat for bacteria and aquatic invertebrates and are, therefore, also an object for ecotoxicological studies. Furthermore, groundwater can act as a transportation pathway connecting different environmental compartments e.g. soils with rivers or oceans. Groundwater thus plays a role in the distribution of contaminants in the environment.

Transport of Contaminants in Aquifers

The movement of a chemical in groundwater is controlled by three processes: advection, dispersion and reaction. Advection is the transport of a chemical in dissolved form together with the groundwater flow. When a chemical is released from a point source into groundwater with a constant flow direction, a plume is forming downstream of the source. The spreading of the chemical is due to dispersion. There are two reasons for this spreading: First, molecular diffusion causes transport of the chemical independently from advection; Second, differences in groundwater velocities at different scales causes mixing of the groundwater (mechanical dispersion) in the direction of groundwater flow but also perpendicular to it. Several process can retard the transport of chemicals or can cause its removal from the system (e.g. degradation). For the mobility of a chemical, the distribution between immobile solid phase and moving liquid phase is of key importance in groundwater (see chapter 3.4). There are several processes which can lead to the degradation of a compound in aquifers. Microbial activity can contribute to the degradation of chemicals but also abiotic reactions can be of importance. For some chemical, redox reactions are relevant which are discussed in the following section.

Redox Reactions in Aquifers

Many elements are redox-sensitive under environmental conditions. This means they occur naturally in different 'redox states'. For example oxidation or reduction of carbon plays a pivotal role in the energy metabolism of living organisms and carbon occurs in oxidation states from +IV in CO_2 (because the two oxygen atoms both count as –II ((because oxygen is more electronegative than carbon)), and the total molecule should balance out) to -IV in CH4 (because each H-atom counts as +I ((because hydrogen is less electronegative than carbon))). Also potentially toxic elements, such as arsenic, are found in nature in different oxidation states. Important oxidation states of arsenic are +V, (e.g. AsO_4^{3-}, arsenate), + III (e.g. AsO^{3-} 3, arsenite), 0 (elemental arsenic or arsenic associated with sulfide as in FeAsS, arsenopyrite). Arsenic can also have negative oxidation states when it forms arsenides such as $FeAs_2$ (löllingite). Bioavailability, toxicity and mobility of redox sensitive elements are usually strongly dependent on their oxidation state. For example, arsenite tends to be more toxic and more mobile than arsenate. For this reason, assessing the redox state of potentially toxic elements is an important element of environmental risk assessment of groundwater.

Organic contaminants can also undergo redox transformations. At the earth surface, when oxygen is present, (photo-)oxidation is an important degradation pathway for organic contaminants. In subsurface environments, when oxygen concentrations are often very low (anoxic conditions), reduction can play an important role in degradation pathways. For example, the reductive dehalogenation of chlorinated hydrocarbons or the reduction of nitroaromatic compounds have been extensively investigated. The reduction of these compounds can be mediated by microorganisms but they can also occur abiotically on solid surfaces present in the subsurface. In any case, reduction of organic contaminants is only possible when the reaction is thermodynamically feasible. For this reason it is necessary to know the redox conditions in, for example, an aquifer.

Quantitative Assessment of Redox Reactions

As the name indicates, redox reactions combine oxidation of one constituent in the system with the reduction of another and, hence, involve electron transfer. The oxidation of arsenite with elemental oxygen to arsenate has following stoichiometry:

$$AsO3-3+0.5O2 \leftrightarrow AsO3-4$$

It is important that the stoichiometries of redox reactions are not only charge- and mass-balanced but also electron-balanced. Here, arsenic releases two electrons when going from oxidation state +III to +V (arsenite becomes oxidized to arsenate) while one oxygen atom takes up the two electrons and goes from oxidation state 0 to -II (elemental oxygen becomes reduced). For this reaction an equilibrium constant can be obtained and based on the activities (or concentrations) of dissolved reactants and products it can be evaluated whether the reaction is in equilibrium or in which direction the reaction is thermodynamically favorable.

When a natural system contains several different redox-active constituents, a large number of possible redox reactions can be formulated and evaluated separately. In this situation it is more convenient to formulate and compare *half reactions*. For examples, the oxidation of arsenite with oxygen can be split up into the reactions of arsenic and oxygen.

$$H_3SO_4 + 2e^- + 2H^+ \leftrightarrow H_3SO_3 + H_2O \quad E \circ h = 0.575V$$

$$O_2 + 4e^- + 4H^+ \leftrightarrow 2H_2O \quad E \circ h = 1.229V$$

Half reactions are typically formulated as reduction reactions (electrons are on the left hand side of the reaction). The Eho is the standard redox potential and represents the electrical potential, which would be measured in a standardized electrochemical cell which contains on one side, H_3AsO_4, H_3AsO_3 and H^+, all with activities of 1 mol l-1, and a solution containing 1 mol l-1 H+ in equilibrium with H_2 gas with a pressure of 1 bar, on the other side.

In natural environments the pH is usually not 0 and the activities of arsenite and arsenate are not 1 mol l-1. The redox potential, Eh under these conditions can be calculated using the Nernst equation:

$$E_h = E_h^\circ + \frac{RT}{zF} \ln \frac{ox}{red} = 0.575\ V + \frac{8.314\ J\ mol^{-1}\ K^{-1}\ 298\ K}{2 * 96485\ mol\ C^{-1}} \ln \frac{\{H_3AsO_4\}\{H^+\}^2}{\{H_3AsO_3\}}$$

where:

- R is the ideal gas constant (8.314 J mol-1 K-1),
- T the temperature in K,
- z is the number of electrons which are transferred in the reaction,
- F the Faraday constant (96485 mol C-1).

In the ratio ox/red, 'ox' represents the activities or pressures of the constituents on the right hand side of the half reaction, whereby the stoichiometric factor becomes the corresponding exponent, while 'red' represents the right hand side of the half reaction.

The redox potentials of different half reactions can be compared:

- The half reaction with the higher redox potential provides the electron acceptor in the thermodynamically favorable redox reaction,
- The half reaction with the lower potential provides the electron donor.
- In other words, it is thermodynamically favorable that the half reaction with the high potential proceeds from left to right and the half reaction with the low potential from right to left.

Redox Conditions in Aquifers

The redox conditions in an aquifer depends on the inherited inventory of oxidants and reductants during the formation of the geological formation and the processes which have been occurring throughout its history. Oxidants and reductants can have entered the aquifer by diffusion or with the infiltrating water and slowly progressing redox reaction can have modified the assemblage of oxidants and reductants. In the absence of (microbial) catalysis redox reactions often have very slow kinetics. Furthermore, due to photosynthesis, redox reactions are not in equilibrium at the earth's surface and the upper part of the underlying subsurface. As a consequence, the redox conditions in an aquifer can usually not be represented in one unique redox potential. This implies that values obtained for groundwaters with electrochemical measurements, e.g. potentiometric measurements using redox electrodes, might be not representative for the redox conditions in the aquifer. Furthermore, relevant half reactions in the aquifer often involve solids (heterogeneous reactions) with low solubility, implying that the concentrations in solution (for example of Fe_3^+) are too low to be detected. Hence, evaluating the redox conditions in subsurface environments is often challenging.

Oxygen concentrations in groundwaters are often virtually zero, as oxygen in infiltrating rain water or entering the subsurface by molecular diffusion is often consumed before it can reach the aquifer. Hence, 'reducing conditions' typically prevail in aquifers. The redox potential measured in a system may reflect the dominant electron acceptors besides oxygen that are present in the system (Figure 1).

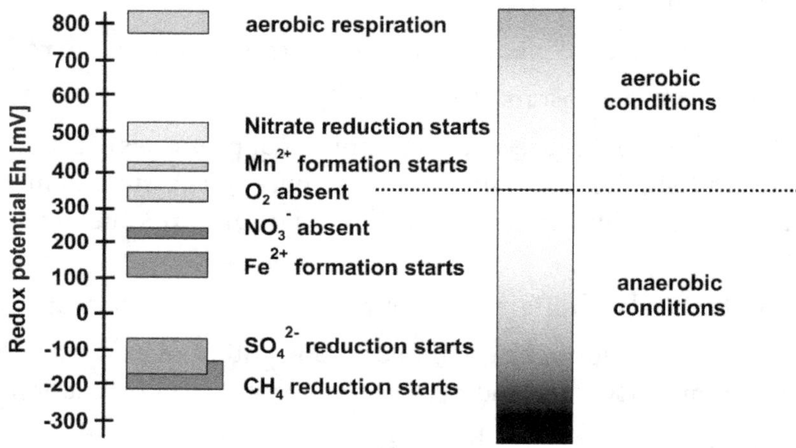

Figure 1. Redox potential scale, ranging from oxic to anoxic conditions. Even when oxygen is absent, the concentrations of available alternative oxidant materials (NO^{3-} and Fe^{2+}) may render a positive redox potential. Methanogenesis (CH4 formation) only initiates at very low negative redox potentials. (Source: Steven Droge)

In sediments or sedimentary rocks, redox reactions after deposition are predominately driven by the oxidation of organic matter, which entered the sediment during its deposition. However, the aquifer might also have received dissolved organic matter via infiltrating water. The oxidation of organic matter is predominately microbially mediated and predominately coupled to the reduction of elemental oxygen (if present). However, when elemental oxygen is depleted, which is usually the case, other electron acceptors are used by microorganisms. Relevant electron acceptors (see Figure 1) in anoxic environments include:

- Nitrate (in dissolved form),
- Mn(IV) (as solid surface),

- Mn(III) (as solid surface),
- Fe(III) (as solid surface),
- Sulphate (in dissolved form).

Nitrate and sulphate can be present in dissolved form while Mn(IV), Mn(III), Fe(III) occur as solids with low solubility. The (hydr)oxide solids of these metals, such as goethite (FeOOH) or manganite (MnOOH) are mostly accessible for microbial reduction while Mn(III) or Fe(III) in silicates can only be partially reduced or are not bioavailable for reduction. When also these electron acceptors run short, methanogenesis can be initiated.

Microorganisms, which reduce sulphate, Mn or Fe(III), can use the products of fermentative organisms. These fermentative organisms produce short-chain fatty acids, such as acetate or lactate, but often also release hydrogen gas. That is, hydrogen concentrations in groundwater reflect a steady state of hydrogen production and consumption, and are typically limited by the rates of production. As a consequence, hydrogen concentrations in groundwater are often at the physiological limit of the consuming organism. The concentrations are just sufficient to allow the organism to conserve energy from oxidizing the hydrogen. This limit increases according to the sequence of electron acceptors (Figure 1): nitrate reduction < Mn reduction < Fe reduction < sulphate reduction < methanogenesis when the corresponding compounds are present in relevant amounts or concentrations. For this reason, concentrations of dissolved hydrogen can be a useful indicator to identify the dominant, anaerobic respiration pathway in an aquifer. For example, one can determine whether sulphate reduction is enabled or methanogenesis has set in. The hydrogen concentrations in the groundwater can also directly be used to assess whether the microbial reduction of metals, metalloids, chlorinated hydrocarbons, nitro aromatic compounds or other organic contaminants is feasible.

The reduction of Fe(III)(hydr)oxides are sulphate leads to the formation of Fe(II) and sulphide, which, in turn, typically results in the precipitation of ferrous solids such as $FeCO_3$ (siderite), FeS (mackinawite) or FeS_2 (pyrite). These Fe(II) containing minerals often play an important role in the abiotic reduction of organic or inorganic contaminants in aquifers. When the composition of the groundwater and the mineral assemblage is known, the Nernst equation can be used to calculate the redox potential of relevant half reactions in the aquifer. These redox potential can be then used for evaluating whether reduction of potentially toxic compounds is possible or not. For example, the half reaction for the reduction of an amorphous ferric iron hydroxide coupled to the precipitation of siderite is given by:

$$Fe(OH)_3(s) + e^- + H^+ + H_2CO_3 \leftrightarrow FeCO_3(s) + H_2O \quad E_h^\circ = 0.579\ V$$

At given pH and carbonic acid concentration, the corresponding redox potential can be calculated using the Nernst equation. This redox potential can be compared to that obtained from the Nernst equation for the reductive dichlorination of tetrachloroethylene ($Cl_2C{=}CCl_2$)

$$Cl_2C = CCl_2 + 2e^- + H^+ \leftrightarrow Cl_2C = CHCl + Cl^- \quad E_h^\circ = 0.70\ V$$

With this approach the feasibility of redox reactions involving potentially organic and inorganic compounds can be evaluated in aquifers. That does, however, not imply that the corresponding reactions also occur within the relevant time scale. For this the kinetics of the reaction have to be known and have been studied for many reactions of potential relevance in aquifer systems. However, the kinetics of redox reactions are not subject of this section.

References

Sparks, D. (2002). Environmental Soil Chemistry, Second Edition, Academic Press, Chapters 5 and 8, ISBN 978-0126564464.

Essington, M.E. (2004). Soil and Water Chemistry: An Integrative Approach, Chapters 7 and 9, CRC Press, ISBN 978-0849312588

3.1.7. Biota

Introduction

Just like soil, water, and air, the organic tissue of living organisms can also be regarded as a compartment of the ecosystem where chemical pollutants can accumulate or can be broken down. The internal concentration in living organisms provide important information on chemical exposure and ultimately determines the environmental risk of pollution, but it is important to understand the key features of tissue that influence chemical partitioning into organisms. Chemical accumulation in the tissue of living organisms is a series of chemical and biological processes, briefly based on:

- Chemical uptake (mostly permeation from bulk media over certain membranes into cells);
- Internal distribution (e.g. via blood flows through organs);
- Metabolism (e.g. biotransformation processes in for instance the liver).
- Excretion (e.g. through urine and feces, but also via gills, sweat, milk, or hairs)

These four processes are the basis of toxicokinetic modeling, and are often summarized as Absorption, Distribution, Metabolism, and Excretion, or "ADME". These ADME processes can strongly vary for different polluting compounds due to the properties of the chemical structure. These ADME processes can also strongly vary for different organisms, because of:

- The physiological characteristics (e.g. having gills, lungs, or roots, availability of specific chemical uptake mechanisms, presence of specific metabolic enzymes, size-related properties like metabolic rate),
- the position in the polluted environment (flying birds or midge larvae living in sediment),
- the interaction with the polluted environment (living in soil or water, food choice, etc.)
- the behaviour in the polluted environment (being sessile or able to move (temporarily) away from a polluted spot).

More details of these toxicokinetic processes are presented in section 4.1 on Toxicokinetics and bioaccumulation. The current module aims to provide a summary of the key features of different tissue components that explain the internal distribution of chemicals (distribution), the different types of contact between pollutants and organisms (exposure-absorption), and temporal changes in physiology that may affect internal exposure (e.g. excretion, which includes examples such as release of POPs via lactation, and increasing POP concentrations during starvation).

Before we discuss how chemicals are taken up into biota, it is important to first define the key chemical properties and the molecular composition of tissue that influence the way chemicals are absorbed from the surrounding environment and distributed throughout an organism.

Absorption-Distribution: Tissue Building Blocks

All organisms are composed of cells, which are composed of a cell membrane, surrounding the largely watery solution filled with inner organelle membranes, protein structures, and DNA/RNA. Prokaryote organisms such as bacteria, but also algae, fungi and plants have reinforced membranes with cell walls to prevent water leaking by high osmotic pressures, and to protect the cell membrane.

Metabolic energy is stored in large molecules such as fatty esters and sugars. Remarkably, for all existing living organism species, these tissue components are mostly structures made out of relatively simple and repetitive molecular building blocks, with minor variations on side chains. See examples in Figure 1. The composition of organs, as a collection of specific cells, in terms of the percentage of lipids, proteins and carbohydrates is important for the overall toxicokinetics of chemicals in the whole organism.

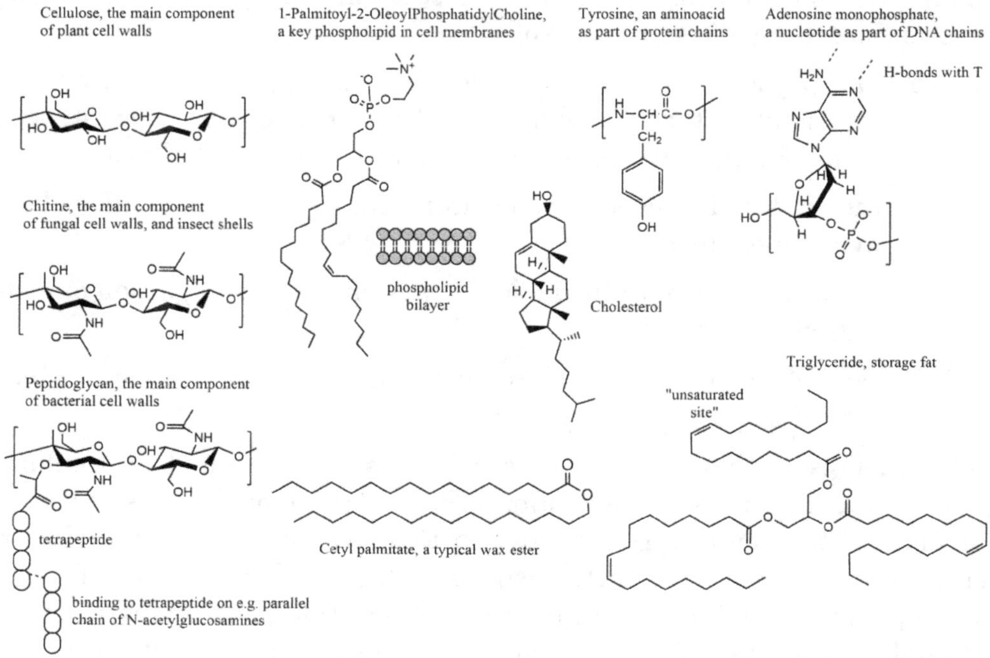

Figure 1. The key molecular components and partial building blocks of living tissue.

Cell walls are mostly made from highly polar polysaccharides, e.g.:

- Cellulose, a polymer of sugary molecules, and chitin in fungi, which is highly polar and therefore permeable for water.
- Peptidoglycan semi-crystal structures surrounding bacteria (90% of the dry weight of Gram-positive bacteria but only 10% of Gram-negative strains), a mixed polymer between N-acetylglucosamine (alike chitine) and short interconnecting 4 or 5 amino acid chains.
- Lignin, a polar (~30% oxygen) but more hydrophobic supra-structure of polymerized phenolic molecules lining the main plant vessels that transport water.

The specific algae group of diatoms have a cell wall composed of biogenic silica (hydrated silicon dioxide), typically as two valves that overlap each other surrounding the unicellular species. Diatoms generate about 20 percent of the oxygen annually produced on the planet, and contribute nearly half of the organic material found in the oceans. With their specific cell wall structure, diatoms take in over 6.7 billion metric tons of silicon each year from the waters in which they live, which creates huge deposits when they die off.

Cell membranes are made up mostly of a phospholipid bilayer, with each phospholipid molecule basically having a polar and ionized headgroup connected to two long alkyl chains (Figure 1 example with POPC type phospholipid). The outer sides of a phospholipid bilayer are hydrophilic (water-loving), the inside is hydrophobic (water-fearing). Ions (inorganic salts, nutrients, metals, strong acids and ionized biomolecules) do not readily permeate through such a membrane passively, and require specific transport proteins that can transport as well as regulate ions in and out of the cell interior. Cholesterol molecules stabilize the fluidity of the membrane bilayers in cells of most organisms, but for example not in most Gram negative bacteria. Dissolved neutral chemicals may passively diffuse through phospholipid bilayers into and out of cells.

Proteins are chains of a variety of amino acids, 21 of which are known to be genetically coded, and of which humans can only produce 12. The other nine must be consumed, and are therefore called essential amino acids (coded H, I, L, K, M, F, T, W, V). Proteins form complex 3 dimensional structures that allow for enzymatic reactions to occur effectively and repeatedly. There are two amino acids with side chains that carry a positive charge at neutral pH: Arginine (pK_a 12) and Lysine (pK_a 10.6), and two amino acids with side chains that carry a negative charge at neutral pH: Aspartic acid (pK_a 3.7) and Glutamic acid (pK_a 4.1). Some amino acids carry typical hydrophobic side chains: amongst others Leucine and Phenylalanine. Cysteine has a thiol (SH) moiety that can form strong connective disulfide interactions with spatially nearby cysteine side groups in the 3D structure. The key blood transport protein albumin, for example, contains about 98 anionic amino acids, and 83 cationic amino acids, and about 35 cysteine residues.

DNA and other genetically encoding chains are composed of 4 different nucleotides that form a double helix of two opposing strands, held together by hydrogen bonds connecting the complementary bases: A and T (or A and U in RNA), and G and C. DNA can be densely packed around histone proteins, and is either or part of the cellular cytoplasm (in prokaryotic species) or separated within a membrane (in eukaryotic species). DNA is not a critical accumulation phase for chemicals, but of course it is a cellular structure where pollutants can strongly impact all kinds of cellular processes when they react with DNA components or affect the structural organisation otherwise.

Storage fat provides for many animals and fruits of plants an important energy reserve, but also insulates warm-blooded animals in cold climates, lubricates joints to move smoothly, and protects organs from shocks (e.g. eyes and kidneys). Seeds and nuts may contain up to 65% (walnuts) of fatty components, which of course provides energy for initial growth, but from which also oil can be pressed. Storage fat in most animals is present in the form of triglycerides, and as such neutral and very hydrophobic phases within tissue. Polyunsaturated fatty acid esters like omega-6 and omega-3 fatty acids are abundant in fish (eicosapentaenoic acid (EPA) and docosahexaenoic acid (DHA)) and in seeds and plants (mostly alpha-linolenic acid (ALA), but algae also contain EPA and DHA). The high intake of algae by fish in aqueous food webs based on algae, results in the high EPA/DHA levels in many fish species, as they mostly cannot make it themselves (https://www.pufachain.eu). Humans can make some EPA and DHA from ALA.

The average composition of living organisms based on the key tissue components lipid, protein, and carbohydrate can range widely, as illustrated in Table 1.

Table 1. Tissue structure composition of the average dry weight of different organisms.

Organism	Lipid % of d.w.	Protein % of d.w.	Carbohydrate % of d.w.
Grass	0.5-4	15-25	60-84
Phytoplankton	20	50	30
Zooplankton	15-35	60-70	10
Oyster	12	55	33
Midge larvae	10	70	20
Army cutworms (moth larvae)	72% of body		
Pike filet	3.7	96.3	0
Lake trout	14.4	85	0
Eel (farmed for 1.5y)	65	~34	1
Deer game meat	10	90	0

Table 2. Estimates on the tissue structure composition of a woman (BW= 60 kg, H = 163 cm, BMI = 22.6 kg/m2), (taken from Goss et al., 2018). Bones are not included.

Organ	Total organ volume (mL)	moisture content	phospholipid % of d.w.	storage lipid % of d.w.	protein % of d.w.
Adipose	22076	26.5%	0.3%	93.6%	6.1%
Brain	1311	80.8%	35.4%	22.6%	42.1%
Gut	1223	81.8%	9.9%	22.0%	68.1%
Heart	343	77.3%	19.1%	17.7%	63.2%
Kidneys	427	82.4%	16.5%	5.9%	77.7%
Liver	1843	79.4%	19.4%	8.0%	72.6%
Lung	1034	94.5%	13.1%	13.7%	73.2%
Muscle	19114	83.7%	2.6%	2.5%	95.0%

Organ	Total organ volume (mL)	moisture content	phospholipid % of d.w.	storage lipid % of d.w.	protein % of d.w.
Skin	3516	71.1%	2.6%	22.9%	74.5%
Spleen	231	83.5%	5.6%	2.4%	92.0%
Gonads	12	83.3%	18.8%	0.0%	81.3%
Blood	4800	83.0%	2.5%	2.4%	95.1%
total	55929	60.2%	1.7%	71.0%	27.3%

Different organs in a single species can also largely differ in their composition, as well as their contribution to the overall body, as shown for a human in Table 2. Most organs have a moisture content >75%, but overall the moisture content is considerably lower, due to the low moisture content of bones and adipose tissue. Adipose tissue is by far the largest repository of lipids, but made up mostly of storage lipid, while the brain is also particularly rich in both lipids, but particularly enriched in phospholipids of cell membranes. Muscles and blood contain relatively high protein content.

Absorption-distribution: Chemical Properties

The influence of chemical structure on the accumulation of chemicals in the biotic compartments is largely dependent on their bioavailability, as discussed in more detail in section 4.1 on Toxicokinetics and bioaccumulation, as well as on the basic binding properties as the results of the chemical's hydrophobicity and volatility (section 3.4 on Partitioning and partitioning constants) and ionization state (section 2.2.6 on Ionogenic organic chemicals). In brief, the more non-polar the composition of a chemical, the more hydrophobic it is, the higher its affinity to partition from dissolved phases (both externally as well as internally) into poorly hydrated tissue phases such as storage fat and cell membranes. For this reason, the main issue with classical organic pollutants such as dioxins, DDT, and PCBs, is often their high hydrophobicity which results in strong accumulation in tissue. Such chemicals often take very long times to excrete from the tissue if they are not made less hydrophobic via biotransformation processes. This leads to foodweb accumulation and specific acute or chronic toxic effects at a certain organism level (section 4.1.6 on Food chain transfer).

Proteins and sugary carbohydrates are mostly comprised of extended series of polar units and thus strongly hydrated, and bind hydrophobic chemicals to a much lower extent. Proteins may have three dimensional pockets that could fit either hydrophilic of hydrophobic chemicals, and as such act as transport proteins in blood throughout the body (transporting fatty acids for example), based on the specific binding affinity. Many protein based receptors are also based on a specific binding affinity, and in many cases this involves (combinations of) polar and electrostatic interactions that also have an optimum three-dimensional fitting space. Volatile chemicals are more abundantly present in the gas phase rather than being dissolved, and are more readily in contact with biota via gas-exchange on the extensive surfaces of lungs of animals and leaves of plants.

In order to be taken up into cells, or into organs, chemicals have to permeate through membranes. For most organic pollutants, the passive diffusion through phospholipid bilayers has an optimum at a certain hydrophobicity. The high accumulation in the membrane ensures desorption into the adjacent cellular solution. It is assumed that ionized chemicals have a passive permeation rates that are either negligible or at least orders of magnitude lower than that of corresponding neutral chemicals. For this reason, all kinds of molecular intra-extracellular gradients can be readily maintained, for example for protons (H^+) or sodium-potassium (Na^+/K^+). The movement of very polar and ionic chemicals can be tightly regulated by transport proteins protruding the membrane bilayer. Specific molecules can be actively excreted from cells (e.g. certain drugs) or reabsorbed (e.g. in the kidneys back into the blood stream). This again is based on three dimensional fitting in the transport pocket and stepwise movement through the protein structure, and costs energy. For acids and bases with a very small fraction of neutral species at physiological pH, the passive permeation over membranes may still be dominated by the neutral species.

Exposure: Contact between Biota and Various Environmental Compartments

There are multiple routes by which chemicals can enter the tissue of biota, for example via respiratory organs, through digestion of contaminated food, or dermal contact. Most animals need to take in a more or less constant flux of oxygen and water, and periodically food, to release nutrients and energy from the food. Of course they also need to release CO_2 (and other chemicals) as waste. Pollutant chemicals are taken up alongside these basic processes, and it depends on the chemical properties and the efficiency of the uptake route how much the organism will take in from these different exposure routes.

Plants need plenty of water and during daytime photosynthesis need CO2, but also require oxygen during the night. High algal densities can deplete the oxygen levels in shallow aquatic systems during the night, and replenish oxygen levels during daytime. Oxygen is plentiful in air (200,000 parts per million in the air), but it is considerably less accessible in water (15 parts per million in cool, flowing water), and often depleted below the first few mm of sediment. To obtain sufficient oxygen, water and food, aquatic organisms have to pass large volumes of waters through their gills. Sediment-dwelling organisms either have hemoglobin to bind oxygen, or constantly pump fresh overlying water through burrows created in sediment, often lined with mucus. Living organisms are thus constantly in contact with water dissolved pollutants, and air breathing organism are readily exposed to air pollutants. To simplify the domain of living organisms as part of this module about the biotic compartment and how they get into contact with chemicals relevant to Environmental Toxicology, they can for example be divided in:

- Which often take in large amounts of water from their surroundings via their roots, driven by the evaporation of water at the leaves and resulting internal flows.
- Water breathing organisms, which pass large amounts of water through gills or gill like structures (tubules or other thin skin structures close to where water is passing) in order to take in enough oxygen and reduce the built up of CO_2. Filter feeders like oysters and mussels, that populate enormous surfaces as reefs or banks, can turnover a huge volume of water on a daily basis, and thus allow dissolved chemicals get in close contact to the outer membranes.
- Air breathing organisms, which can effectively exchange large quantities of volatile and gaseous chemicals with the air (oxygen but also organic compounds), but typically take in less/non-volatile chemicals via food and require active excretion and metabolism for the emission of less/ non-volatile chemicals.

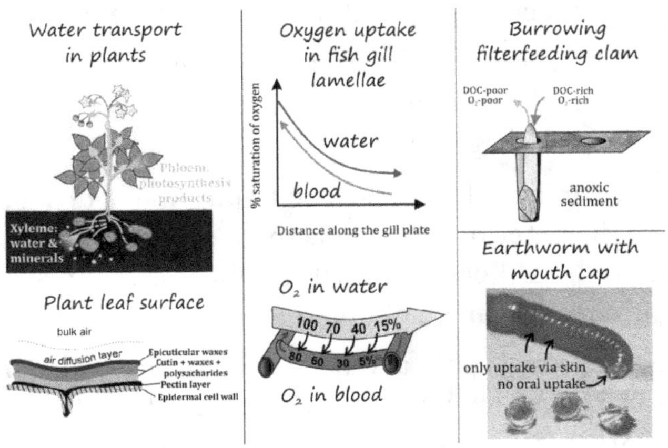

Figure 2. Left panel: Plant water transport (adapted from https://en.wikipedia.org/wiki/Xylem), a schematic view of the waxy upper layers on leaves (adapted from Moeckel et al. 2008)). Middle panel: water flushing along gill lamellae replenishes oxygen in blood. Top Right panel: Many benthic organisms living in anoxic sediments require overlying oxygen rich water, and feed on dissolved particles. Lower Right Panel: Earthworms take up oxygen via diffusion through their skin, but also metals and organic pollutants, as shown in studies where the mouth part has been glued shut during exposure (adapted from Vijver *et al.*, 2003).

Plants

Nearly all plants have roots below ground, a sturdy structure of stem and branches above ground, and leaves. Along with soil pore water, soluble chemicals are readily transported from roots of the plant in the internal circulation stream through xylem cells, which are lined with water impenetrable lignin (see Figure 2). Moderately hydrophobic chemicals (Kow of 1-1000) are rapidly transported from roots to shoots to leaves, while hydrophobic chemicals may be strongly retained on the membranes and cell walls and mostly accumulate in root sections, limiting transport to above-ground plant tissues. Roots may also actively release considerable quantities of chemicals to influence the immediate surrounding media of the roots (rhizosphere), e.g. to stimulate microbial processes or pH in order to release nutrients. These plant root 'exudates' can be ions, small acids, amino acids, sterols, etc. Chemicals that enter plants via leaves, such as pesticides or semi-volatile organic pollutants, can be redistributed to other plant parts via the phloem streams.

The transport through xylem up to higher plant tissues occurs via capillary forces and is enhanced by three passive phenomena:

- The high sugar content in the phloem causing osmotic pressure to attract water from other parts.
- The evaporation of water creates surface tension on thousands of cells that pulls water from the soil through the xylem system.
- The osmotic pressure of root cells compared to soil pore water. Root pressure is highest in the morning before the stomata open and allow transpiration to begin.

As a result of the capillary forces needed to pull water up against gravity, and a certain maximum diameter of the vessels to do so, there is a maximum possible plant height of 122-130 m (Koch *et al.*, 2004) which compares to Redwood trees (Sequoias) reaching a maximum height of 113 m.

Most leaves are covered with a waxy layer, to prevent damage and water evaporation. This wax layer may be 0.3-4.6 µm thick (Moeckel et al., 2008). Large forests provide enormous hydrophobic surfaces to which semi-volatile organic chemicals (SVOC) can bind out of air, which influences the global distribution of chemicals such as PCBs. Partitioning of SVOCs on the vegetation of extended grasslands contaminates the base of the food chain, as well as agricultural and cattle sectors used by humans. The grass/corn-cattle-milk/beef food chain accounts for the largest portion of background exposure of the European and North American population to many persistent SVOCs. The absorption rate of chemicals on the leaves often also depends on the air boundary layer surrounding leaves, which limits diffusion into the leave surfaces. Of course, all kinds of other factors such as wind speed, canopy formation and cuticle thickness also control exchange between leaves and gas phase (see also section 3.1.2 on the Atmosphere). Tiny openings or pores on the lower side of the leaves, called stomata, allow furthermore for gas exchange. In warm conditions, stomata can close to prevent water evaporation, but gas exchange is needed in many plant types to allow for CO_2 to be metabolized and the release of O2 that is produced. The waxy layer on leaves can trap gaseous organic chemicals. Many plants, like coniferous trees, produce resins to provide effective defense against insects and diseases, and these resins release large amounts and structurally highly diverse organic volatiles such as terpenes and isoprenes (Michelozzi, 1999). These plant-produced volatiles can even contribute to ozone formation. Plants thus accumulate chemicals from their environment, but also release chemicals into the environment. It thus also matters for the exposure of grazing organisms to certain types of pollutants whether they eat roots, shoots, leaves, seeds or fruits of plants living in contaminated environments.

Organisms Using Dissolved Oxygen

The 'gill' movements of water breathers create a constant flux of chemicals dissolved in bulk water along outer cell membranes (or mucus layers surrounding cell membranes) of gills. A 1 kg rainbow trout fish ventilates about 160 mL/min, so 230 L/day (Consoer et al., 2014). The total gill surface area in a fish depends on species behaviour and weight (active large fish require a lot of oxygen), and equals to about 1-6 cm2 per g fish (Palzenberger & Pohla, 1992). For a 1 kg fish of ~20 cm length, the ~1000 cm2 gill area compares to a ~500 cm2 outer body surface. This results in an effective partitioning of chemicals between water and cell membranes. Within the gills, the cells are in close contact with the blood system of the organism, and the build-up of chemical concentrations in the outer cells provides an effective exchange with the blood stream (or other internal fluids, Figure 2) flushing along that redistribute chemicals to the inner organs. The reverse equally occurs: chemicals dissolved in blood stream coming from organs will also rapidly exchange with bulk external water if concentrations are lower.

Of course, many pollutants can also enter water breathing organisms via food, but the gills-water exchange is very effective in controlling the distribution of chemicals. The salinity of water plays a strong role in the need of water breathing organisms to "drink" water, and hence take in contaminants via this route.

A single adult oyster can cleanse about 200 liters of water per day (https://www.cbf.org/about-the-bay/more-than-just-the-bay/chesapeake-wildlife/eastern-oysters/oyster-fact-sheet.html). Plans to re-populate the harbour of New York with 1 billion oysters on artificial substrates can have enormous impacts on chemical redistribution. A single 2 cm zebra mussel (Dreissena polymorpha) that inhabits the shallow Lake IJsselmeer (6.05x1012 L) in can filter about 1 L per day, and the high densities of these and related species in this fresh water lake can turn over the lake volume once or twice per month (Reeders et al., 1989).

BOX 1. Osmoregulation (MSc level)

Most aquatic vertebrate animals are osmoregulators: their cells contain a concentration of solutes that is different than the water around them. Fish living in freshwater typically have a cellular osmotic level (300 millOsmoles per Liter, mOsm/L) that is higher than the bulk fresh water (~20-40 mOsm/L), so a lot of water flows passively via the gills (not via the skin) into the tissue of fish. They are thus constantly taking in water (water molecules only) via the gills, which needs to be controlled, e.g. by strongly diluting the urine. They do also take in some water in their gastro-intestinal tract (GIT). Marine fish have similar cellular osmotic levels as freshwater fish, but the salty water (1000 mOsm/L) causes water to move out of the gill tissue through the linings of the fish's gills by osmosis, which needs to be replenished by active intake of salty water, and separate excretion of the salts. Most invertebrate organisms in oceans have an internal overall concentration of dissolved compounds comparable to the water they live in, so that they don't suffer from strong osmotic pressures on their soft tissue (osmoconformers).

Even many soil organism are constantly in contact with wet soil surfaces, and contact between soil pore water and the outer surfaces (gills/soft skin areas) dominates the routes of chemical exchange for many chemicals. Earthworms, for example, do not have lungs and they exchange oxygen through their skin. Earthworms eat bacteria and fungi that grow on dead and decomposing organic matter, and thus act as major organic matter decomposers and recycling of nutrients. Earthworms dramatically alter soil structure, water movement, nutrient dynamics, and plant growth. It is estimated that earthworms turn over the top 15 cm of soil in ten to twenty years (LINK), and so they also are able to mix surface bound pollution into a substantial soil layer.

In terms of biomass, earthworms dominate the world of soil invertebrates, including arthropods. In order to better understand how much contamination earthworms take in via food or via their skin, several studies have used earthworms in exposure tests with part of the organisms having their mouth parts sealed with surgical glue (Vijver et al., 2003). Uptake rates of the metals Cd, Cu and Pb in sealed and unsealed earthworms exposed to two contaminated field soils were similar (Vijver et al., 2003), indicating main uptake through the skin of the worms. The uptake rates as well as the maximum accumulation level for several organic contaminants from artificially contaminated soil were also comparable between sealed and non-sealed worms (Jager et al., 2003). The dermal route is thus a highly important uptake route for organic chemicals too. Dermal uptake by soil organisms is generally from the pool of chemicals in the soil pore water, hence the distribution of chemicals between solid particles and organic materials in the soil and the soil pore water is extremely important in driving the dermal uptake of chemicals by earthworms (See section 3.4 on Partitioning and partitioning constants, section 3.5 on Metal speciation and 3.6 on Availability and bioavailability).

Air Breathing Organisms

Air breathing organisms typically take in less/non-volatile chemicals via food and require active excretion and metabolism for the elimination of these chemicals via e.g. feces and urine. Dermal uptake is generally assumed to be negligible, while the intake of chemicals via air particles can be relatively high, for example of pollutants present in house dust, or contaminants on aerosols. The excretion via air particles is clearly not a dominant route. The chemicals in the food matrix inside the gastro-intestinal tract often first need to be fully dissolved in gut fluids, before they can pass the mucus layers and membranes lining the gastrointestinal tract and enter blood streams for redistribution. However, 'endocytosis' may also result in the uptake of (small) particulate chemicals in cells by first completely surrounding the particle by the membrane, after which the encapsulating membrane buds buds off inside the cell and forms a vesicle. Notwithstanding this endocytosis, chemical fractions of pollutants strongly sorbed to non-digestible parts may not always automatically be taken up from food. Grazing animals typically require microbial conversion in their gut to digest plant material like cellulose and lignin into chemical components that can be taken up as energy source.

Many aquatic foodwebs are structured such that they begin with aquatic plants being eaten by water breathing organisms, with air breathing marine animals or birds on top of the food chain. These air breathing top predators take in pollutants largely through their diet, but lack the effective blood-membrane-water exchange through gills. The blood-membrane-air partitioning in lungs is far less effective in removing chemicals via passive partitioning. For this reason, many top predators of foodwebs have the highest concentrations of pollutants. The chemical distribution in foodwebs will be discussed in more detail in section 4.1.6.

References

Palzenberger & Pohla 1992, Reviews in Fish Biology and Fisheries, 2, 187-216.

Jager, T.; Fleuren, R.H.L.J.; Hogendoorn, E.A.; de Korte, G. 2003. Elucidating the routes of exposure for organic chemicals in the earthworm, Eisenia andrei (Oligochaeta). Environ. Sci. Technol. 37 (15), 3399-3404

Koch et al. 2004, Nature (428) 851–854.

Moeckel et al. 2008, Environ. Sci. Technol. 42, 100–105.

Michelozzi 1999, Defensive roles of terpenoid mixtures in conifers, Acta Botanica Gallica 146 (1), 73-84.

Consoer et al. 2014, Aquatic Toxicology 156, 65–73.

Reeders et al. 1989, Freshwater Biology 22 (1), 133-141.

Vijver et al. 2003, Soil Biology and Biochemistry 35, 125-132.

Goss et al. 2018, Chemosphere 199, 174-181.

3.2. SOURCES OF CHEMICALS

Chemicals can be released into the environment in many different ways. In the professional field, this release is called the emission of the chemical, and the origin of the release is referred to as the emission source. The strength and nature of the emission source(s) are important determinants of the ultimate environmental exposure and thus of the resulting risk. This section explains the most important characteristics of emission sources and familiarizes you with the most common terms used in the field of emission assessment. It starts with a brief introduction on the origin of pollutants in the environment, followed by an explanation of the relevance of emission assessment, how an emission can be characterized, and how data on emissions can be gathered or estimated.

Origin of Pollutants

Pollutants in the environment can originate from different processes. Within this context, we here distinguish between three types of chemicals:

- *Natural chemicals* that are naturally present in the environment, like (heavy) metals and natural toxins, and that can either be released into the environment by natural processes (e.g. the eruption of a volcano) or by human-induced processes (e.g. resource extraction and subsequent use and dispersal);
- *Synthetic chemicals* that are intentionally produced and used by society because of their useful characteristics, e.g. (most) pharmaceuticals, pesticides and plastics;
- Chemicals that are *unintentional byproducts* of human activities and production processes, e.g. dioxins and disinfection byproducts.

The latter category can overlap with the first, since the reaction products of natural processes, such as many combustion processes, can be considered natural chemicals. Polycyclic aromatic hydrocarbons (PAHs), for example, can be released from natural (e.g. a forest fire) as well as human-induced processes (e.g. a power plant). This emphasizes the role of the emission process in defining an environmental pollutant. When human activities are involved in either the production or the release of a chemical into the environment, this chemical is considered to be an environmental pollutant. Some synthetic chemicals are also naturally present in the environment as this is a specific field of research in organic chemistry, i.e. the chemical synthesis of natural products.

The Relevance of Emission Assessment

Emission assessment of chemicals is the process of characterizing the emission of a chemical into the environment. Knowledge on the emission can be relevant for different purposes. The most obvious purpose is to assess the exposure and risks of a chemical in the vicinity of the emission source. This is typically done when a facility requires an environmental permit to operate, e.g. a discharge permit for surface water or a permit that involves the emission of pollutants into air from a smoke stack. Such assessments are typically performed locally.

At a higher scale level, e.g. national or global, one might be interested in all emissions of a certain compound into the environment. One should then map all sources through which the chemical can be released into the environment. For synthetic chemicals, this implies the mapping of all emissions throughout the life cycle of the chemical. This life cycle is typically divided into three phases: production, use and waste (Figure 1). Between the production and use of a chemical there may be various intermediate steps, such as the uptake of the chemical in a formulation or a product. And after a chemical – or the product it is contained in – has been used, it may be recycled. The life cycle of chemicals can be illustrated with the simple example of pharmaceuticals. These can be released into the environment during: (1) their production process, e.g. the effluent of a production plant that is being discharged into a nearby river, (2) their use, e.g. the excretion of the parent compound via urine and feces into the sewer system and subsequently the environment, or (3) their waste phase, e.g. when unused pharmaceuticals are flushed through the toilet or dumped in a dustbin and end up with the solid waste in a landfill.

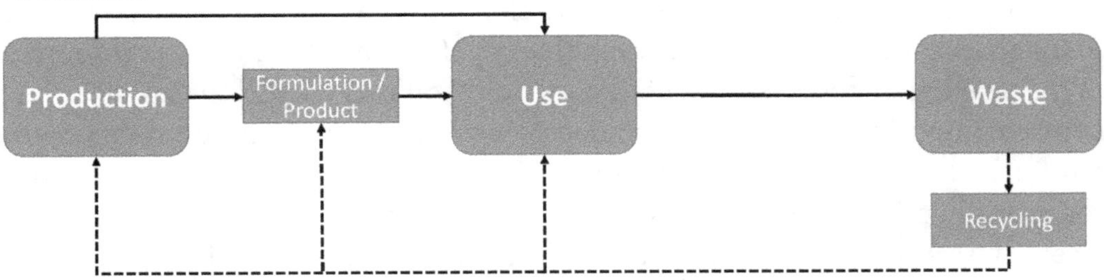

Figure 1. The three main life cycle phases of a chemical: production, use and waste. After production, chemicals can be applied in a formulation or product. After the chemical (or product) becomes waste, it can be recycled to be used again in production, the formulation/product or use phase.

Instead of focusing on the life cycle of an individual chemical, it is more common in environmental assessments to focus on the life cycle of products and services. The life cycle of products and services has an extra phase, i.e. resource extraction. The focus on products or services is particularly useful when one wants to select the most environmentally friendly option from a number of alternatives, e.g. the choice between putting milk in glass or carton. This requires that not only emissions of chemicals are being included in the life cycle assessment, but also other environmental impacts such as the use of non-renewable resources, land use, the emission of greenhouse gases and disturbance by noise or odor. Similar techniques to assess and compare the environmental impacts of human activities include material flow analysis, input/output analysis and environmental impact assessment. The quantification of chemical emissions into the environment is an important step in all these assessment techniques.

Characteristics of The Emission Source

Emission sources can be characterized based on their properties. An important distinction that is often made is the distinction between point sources and diffuse sources. Point sources are emission sources that are relatively few in number and emit relatively large quantities of chemicals. The smoke stacks of power plants and the discharge pipes of wastewater treatment plants (WWTPs) are typical examples of point sources. *Diffuse sources* are many in number and emit relatively small amounts of chemicals. Exhaust emissions from cars and volatilization of chemicals from paints are typically considered diffuse emissions. The distinction between point sources and diffuse sources can sometimes be a bit arbitrary and is particularly relevant within regulatory contexts since point sources are generally more easy to control than diffuse sources.

Another important characteristic of emission sources is the compartment to which the chemical is being emitted, in combination with the matrix in which the chemicals are contained. Two important emission types are chemicals in (waste) water discharged into surface waters and chemicals in (hot) air released through a smoke stack. Other common entry pathways of chemicals into the environment are the spaying of pesticides (emission into air, soil and water), the application of manure containing veterinary medicines, the dumping of polluted soils (in soils or water), the dispersal of polluted sediments, and leaching of chemicals from products. Chemicals emitted to air, and to a lesser extent also water, will typically disperse faster into the environment than chemicals emitted to soils. An important aspect that influences dispersal is whether the chemical is dissolved in the matrix or bound to a phase in the matrix, like organic matter, suspended matter or soil particles. The fate of chemicals in the environment is further discussed in Sections 3.3 and 3.4.

The temporal dimension of the emission source is another important characteristic. Distinction is often made between *continuous and intermittent sources*. Wastewater treatment plants (WWTPs) and power plants are typical examples of continuous sources, whereas the application of pesticides is a typical example of an intermittent emission source. The strength of an emission source may vary over time. Distinction can be made between sources with: (1) a constant emission, (2) a regularly fluctuating emission and (3) a irregularly fluctuating emission. For example, WWTPs typically have a continuous emission, but the amount of a chemical in the WWTP effluent may show a distinct regular pattern over 24 hours, reflecting the diurnal and nocturnal activities of people. Production plants that only operate during the day typically show a block pattern, whereas pesticide emissions typically follow a more irregular pattern fluctuating with the season and the emergence of pest species. Irregular emissions such as from pesticides are typically characterized by peak emissions, i.e. the release of relatively large amounts with a relatively short time frame. Other typical examples of peak emissions are the release of chemicals after industrial accidents or intense rain events, e.g. pesticide runoff from agricultural fields after a long period of drought or combined sewer overflows (CSOs).

Emission Data

Considering the importance of emission assessment for assessing the environmental impacts of human activities, it is not surprising that a lot of effort is put in the quantification of emission sources. Emission sources can be quantified in different ways. An important distinction is that between *measurement* and *estimation.* The continuous measurement of an emission source is also referred to as *monitoring.* Measurement often involves the separate determination of two dimensions of the emission, i.e. (1) the concentration of the chemical in the matrix that is being emitted, and (2) the flow of the matrix into the environment, e.g. the volume of wastewater or polluted air released per unit of time. The emission load (i.e. mass of chemical released per unit of time) is subsequently calculated by multiplying the concentration in the matrix by the flow of the matrix.

Measurement is often costly and takes a lot of time. It is therefore not surprising that approaches have been developed to estimate emissions. These estimations are often in essence based on measurement data, but these are then generalized or extrapolated to come to more large scale emission estimations. For example, measurements on the exhaust emissions of a few cars can be extrapolated to an entire country or continent if you know the number of cars. A rather coarse approach that is widely used for emission estimation is the use of emission factors. An *emission factor* quantifies the fraction of a chemical being used and that ultimately reaches the environment. It is often a conservative value that is based a worst case interpretation of the available measurement data or data on the processes involved in the release of the chemical. A related but more detailed approach is to estimate the emission of a chemical based on proxies such as the amount produced, sold or used in combination with specific data on the release process. Pharmaceuticals can again serve as a good example. If you know the amount of a pharmaceutical that is being sold in a particular country, you can calculate the average per capita use. You can then estimate the amount of pharmaceutical that is being discharged by a particular WWTP if you know: (1) the number of people connected to the WWTP; (2) the fraction of the pharmaceutical that is being excreted by the patient into the sewer system through urine and feces, and (3) how much of the compound is being degraded in the WWTP. You can even further refine this estimation by accounting for (1) demographic characteristics of the population since older people tend to use more pharmaceuticals than young people, and (2) the fractions that are not used by the patient and are either: (a) flushed through the toilet, (b) dumped in the dustbin, or, preferably, (c) returned to the pharmacy.

Emission data can be a valuable source of information for risk assessors. Data gathered locally may be relevant to obtain a picture of national or even global emissions. This insight has led authorities to set up databases for the registration of emissions. Examples of such databases include:

- The European Pollutant Release and Transfer Register (E-PRTR), containing data reported by EU member states on releases to air, water and land as well as the transfers of pollutants in waste water for 91 substances and across 65 industrial sub-sectors, and the transfer of waste from these industrial facilities;
- Waterbase, containing data on the status and quality of Europe's rivers, lakes, groundwater bodies and transitional, coastal and marine waters, on the quantity of Europe's water resources, and on the emissions to surface waters from point and diffuse sources of pollution;
- The Toxics Release Inventory of the United States Environmental Protection Agency, containing data on how much of each chemical is released to the environment and/or managed through recycling, energy recovery and treatment as reported by different industry sectors;
- The data from California's pesticide use reporting (PUR) program, containing data on pesticide use in California;
- The Dutch Watson database (in Dutch), reporting measurements of chemicals in the influent and effluent of Dutch WWTPs.

3.3. PATHWAYS AND PROCESSES DETERMINING CHEMICAL FATE

Characterizing 'fate'

Chemicals can escape during all steps of their life cycle, e.g. manufacturing, processing, use, or disposal. Release of chemicals into the environment necessarily leads to exposure of ecosystems, populations, and organisms including man. Exposure assessment science seeks to analyze, characterize, understand and (quantitatively) describe the pathways and processes that link releases to exposure. Chemicals in the environment undergo various transport, transfer and degradation processes, which can be described and quantified in terms of loss rates, i.e. the rates at which chemicals are lost from the environmental compartment into which they are emitted or transferred from adjacent compartments. Exposure assessment science aims to capture the 'environmental fate' of chemicals in process descriptions that can be used in mass balance modeling, using mathematical expressions borrowed from thermodynamic laws and chemical reaction kinetics (Trapp and Matthies, 1998).

The 'fate' of a chemical in the environment can be viewed of as the net result of a suite of transport, transfer and degradation processes (see Section 3.4 on partitioning and partitioning constants, Section 3.6 on availability and bioavailability, Section 3.7 on degradation) that start to act on the chemical directly after its emission (see Section 3.2 on sources of emission) and during the subsequent environmental distribution. Environmental fate modeling (see Section 3.8 on multimedia mass balance modelling) builds on this knowledge by implementing the various degradation, transfer and transport processes derived in exposure assessment science in mathematical models that simulate 'fate of chemicals in the environment'.

First-order Kinetics

In chemical reaction kinetics, the amount of chemical in a 'system' (for instance, a volume of surface water) is described by mass balance equations of the kind:

$$\frac{dm}{dt} = i - k\,m \qquad \text{(eq. 1)}$$

where $\frac{dm}{dt}$ is the rate of change (kg.s-1) of the mass (kg) of chemical in the system over time t (s), i is the input rate (kg.s-1) and k (s-1) is the reaction rate constant. Mathematically, this equation is a first-order differential equation in m, meaning that the loss rate of mass from the systemis proportional to the first power of m. Equation 1 is widely applied in description and characterization of environmental fate processes: environmental fate processes generally obey first-order kinetics, and can generally be characterized by a first-order reaction rate constant k1st:

$$\frac{dm}{dt} = -k^{1st} m \qquad \text{(eq. 2)}$$

Such loss rated equations can also be formulated in integral format, which is obtained by integration of equation (2) over time t with initial mass $m_0 = m(0)$:

$$m_t = m_0\, e^{-k^{1st}\, t} \qquad \text{(eq. 3)}$$

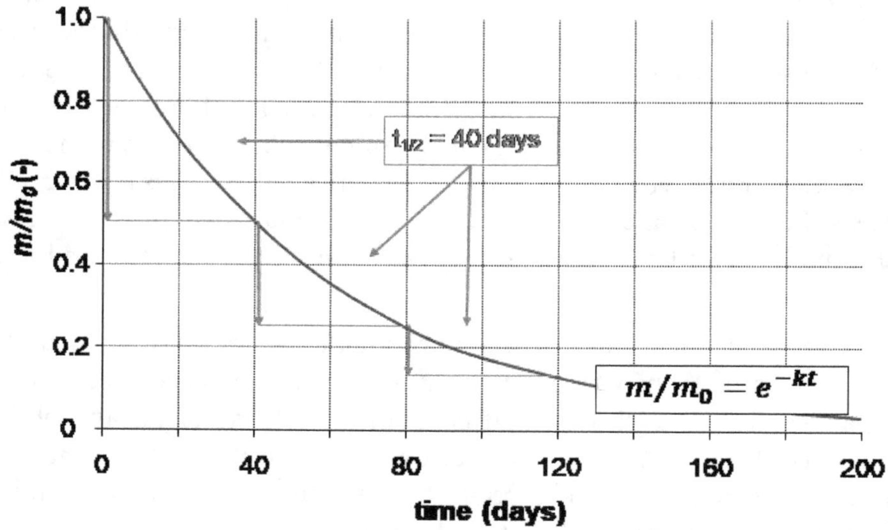

Figure 1. Graphical representation of equation 3. Decrease of relative mass of a chemical in an environmental compartment would follow the blue curve when the loss process is given by a first-order differential loss equation. Loss processes that obey first-order kinetics have constant half-lives (here time $t_{1/2}$ =40 days).

As shown in Figure 1, first-order loss processes are expected to result in exponential decrease of mass from which concentration can be calculated by dividing m with the compartment volume. Using the value $t_{1/2}$ for t in equation 3, it follows directly that the value of $t_{1/2}$ is inversely proportional to the first-order loss rate constant k^{1st}:

$$\frac{m_t}{m_0} = \frac{1}{2} = e^{-k^{1st} t_{1/2}} \rightarrow ln(\frac{1}{2}) = -k^{1st} t_{1/2} \rightarrow t_{1/2} = \frac{ln(2)}{k^{1st}} \quad \text{(eq. 4)}$$

Which shows that half-life time constant, i.e. independent of the concentration of the chemical considered. This is the case for all environmental loss processes that obey first-order kinetics. First-order loss processes can therefore be sufficiently characterized by the time required for disappearance of 50% of the amount originally present.

The disappearance time DT50 is often used in environmental regulation but is only identically with the half-life if the loss process is of first order. Note that the silent assumption of constancy of half-life implies that the process considered is assumed to obey first-order kinetics.

Abiotic Chemical Reactions

Occurrence of true first-order reaction kinetics in chemistry is rare (see Section 3.7 on degradation). It occurs only when substances degrade spontaneously, without interaction with other chemicals. A good example is radio-active decay of elements, with a reaction rate proportional to the (first power of) the concentration (mass) of the decaying element, as in equation 3.

Most chemical reactions between two substances are of second order:

$$\frac{dm}{dt} = -k^{2nd}\, m_1\, m_2 \qquad \text{(eq. 5)}$$

or, when a chemical reacts with itself:

$$\frac{dm}{dt} = -k^{2nd}\, m_1^2 \qquad \text{(eq. 6)}$$

Because the reaction rate is proportional to the concentrations (masses) of both of the two reactants. It follows directly from equation 2. As the concentrations (masses) of both reactants decrease as a result of the reaction taking place, the reaction rate decreases during the reaction, more rapidly so at high initial concentrations. When second-order kinetics applies, half-life is not constant, but increases with ongoing reaction, when concentrations decrease. In principle, this is the case for most chemical reactions, in which the chemical considered is transformed into something else by reaction with a transforming chemical agent.

In the environment, the availability of second reactant (transforming agent) is usually in excess, so that its concentration remains nearly unaffected by the ongoing transformation reaction. This is the case, for oxidation (reaction with oxygen) and hydrolysis (reaction with water). In these cases, the rate of reaction decreases with the decreasing concentration of the first chemical only:

$$\frac{dm}{dt} = -(k^{2nd}\, m_2)m_1 = k^{1st}\, m_1 \qquad \text{(eq. 7)}$$

And reaction kinetics become practically first-order: so-called *pseudo* first-order kinetics. Pseudo first-order kinetics of chemical transformation processes is very common in the environment.

Biotic Chemical Reaction

Chemical reactions in the biosphere are often catalyzed by enzymes. This type of reaction is saturable and the kinetics can be described by the Michael-Menten kinetic model for single substrate reactions. At low concentrations, there is no effect of saturation of the enzyme and the reaction can be assumed to follow (pseudo) first order kinetics. At concentrations high enough to saturate the enzyme, the rate of reaction is independent of the concentrations (masses) of the reactants, thus constant in time during the reaction, and the reaction obeys zero-order kinetics. This is true for catalysis, where the reaction rate depends only on the availability of catalyst (usually the reactive surface area):

$$\frac{dm}{dt} = -k^{zero}\, m^0 = \text{constant.} \qquad \text{(eq. 8)}$$

One could say that the rate is proportional to the zero-th power of the mass of reactant present. In case of zero-order kinetics, the half-life times are longer for greater initial concentrations of chemical. An example of zero-order reaction kinetics is the transformation of alcohol (ethanol) in the liver. It has been worked out theoretically and experimentally that human livers remove alcohol from the blood at a constant rate, regardless the amount of alcohol consumed.

Microbial Degradation

Microbial degradation (often referred to as biodegradation) is a special case of biotic transformation kinetics. Although this is an enzymatically catalysed process, the microbial transformation process can be viewed of as the result of the encounter of molecules of chemical with microbial cells, which should result in apparent second-order kinetics (first order with respect to the number of microbial cells present, and first order with respect to the mass of chemical present):

$$\frac{dm}{dt} = -(k^{2nd}\, m_{bio})m = -k_{deg}\, m \qquad \text{(eq. 9)}$$

where mbio stands for the concentration (mass) of active bacteria present in natural surface water, and k_{deg} represents a pseudo-first-order degradation rate constant.

Advective and Dispersive Transport

Chemicals can be moved from one local point to another by wind and water currents. Advection means transport along the current axis whereas dispersion is the process of turbulent mixing in all directions. Advective processes are driven by external forces such as wind and water velocity, or gravity such as rain fall and leaching in soil. In most exposure models these processes are described in a simplified manner, e.g. the dispersive air plume model. An example of a first-order advective loss process is the outflow of a chemical from a lake:

$$\frac{dm}{dt} = -\frac{Q}{V}m = -k_{adv}\, m \qquad \text{(eq. 10)}$$

where Q stands for the flow rate of lake water [m^3/s] and V is the lake volume [m^3]. Q/V is known as the renewal rate constant kadv of the transport medium, here water. More sophisticated hydrological, atmospheric, or soil leaching models consider detailed spatial and temporal resolution, which require much more data and higher mathematical computing effort (see sections 3.1.2 and 3.8.1).

Transfer and Partitioning

Due to Fick's first law the rate of transfer through an interface between two media (e.g. water and air, or water and sediment) is proportional to the concentration difference of the chemical in the two media (see section 3.4 on partitioning, and

Schwarzenbach et al., 2017 for further reading). As long as the concentration in one media is higher than in the other, the more molecules are likely to pass through the interface. Examples are volatilization of chemicals from water (to air) and gas absorption from air (to water or soil), adsorption from water (to sediments, suspended solids and biota) and desorption from sediments and other solid surfaces.

When two environmental media are in direct contact, (first-order) transfer can take place in two directions, in the case of water and air by volatilization and gas absorption: Each at a rate proportional to the concentration of chemical in the medium of origin and each with a (first-order) rate constant characteristic of the physical properties of the chemical and of the nature of the interface (area, roughness). This is known as physical intermedia partitioning (see section 3.4 on partitioning), usually represented by a chemical reaction formula:

$$[M]_{water} \underset{\leftrightarrow}{\overset{exchange}{\longrightarrow}} [M]_{air} \qquad \text{(eq. 11)}$$

where $[M]$ stands for a (mass) concentration (unit mass per unit volume) and the double arrow represents forward and reverse transport. Intermedia partitioning proceeds spontaneously until the two media have come to thermodynamic equilibrium. In the state of equilibrium, forward and backward rates (here: volatilization from water to air and gas absorption from air to water) have become equal. At equilibrium, the total (Gibbs free) energy of the system has reached a minimum: the system has come to rest, so that

$$k_{abs}[M]_{air} = k_{volat}[M]_{water} \rightarrow K_{AW} = \frac{[M]_{air}}{[M]_{water}} = \frac{k_{volat}}{k_{abs}} \qquad \text{(eq. 12)}$$

and the ratio of concentrations of the chemical in the two media has reached its (thermodynamic) equilibrium value, called equilibrium constant or partition coefficient (see section 3.4 on partitioning).

Challenge

Challenge to environmental chemists is to describe and characterize the various processes of chemical and microbial degradation and transformation, of intra-media transport and intermedia transfer rate constants and of equilibrium constants, in terms of (i) physical and chemical properties of the chemicals considered and (ii) of the properties of the environmental media.

References

Schwarzenbach, R.P., Gschwend, P.M., Imboden, D.M. (2017). Environmental Organic Chemistry, Third Edition, Wiley, ISBN 978-1-118-76723-8.

Trapp, S., Matthies, M. (1998). Chemodynamics and Environmental Modeling. An Introduction. Springer, Heidelberg, ISBN 3-540-63096-1.

3.4. PARTITIONING AND PARTITIONING CONSTANTS

3.4.1. Relevant Chemical Properties

Introduction

Different processes affect the fate of a chemical in the environment. In addition to the transfer and exchange between compartments (air-water-sediment/soil-biota), also degradation determines the concentration in each of these compartments (Figure 1).

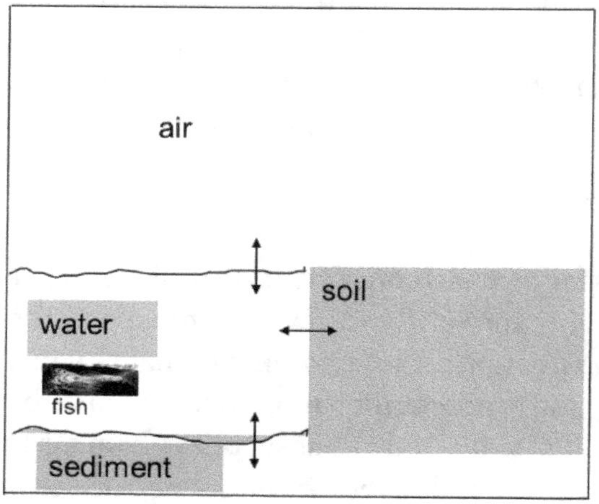

Figure 1. Environmental fate: exchange between compartments and degradation affect the concentration in each compartment.

Some of these processes are discussed in other sections (see sections on Sorption and Environmental degradation of chemicals). Some chemicals will easily evaporate from water to air, while others remain mainly in the aqueous phase or sorb to sediment and accumulate into biota.

These differences are related to only a few basic properties:

- Hydrophobicity (tendency of a substance to escape the aqueous phase)
- Volatility (tendency of a substance to vaporize)
- Degree of ionization

Hydrophobicity

Hydrophobicity means fear (phobic) of water (hydro). A hydrophobic chemical prefers to "escape from the aqueous phase" or in other words "it does not like to dissolve in water". Water molecules are tightly bound to each other via hydrogen bonds. For a chemical to dissolve in water, a cavity should be formed in the aqueous phase (Figure 2) and this will cost energy.

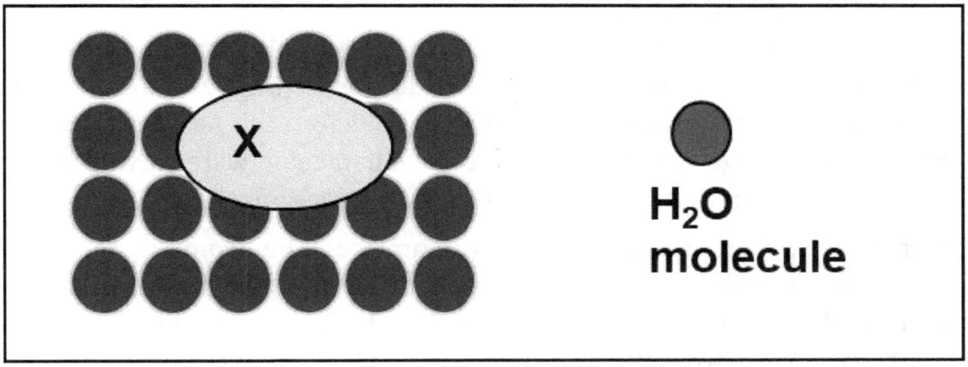

Figure 2. The formation of a cavity in water for chemical X.

Hydrophobicity mainly depends on two molecular properties:

- Molecular size
- Polarity / ability to interact with water molecules, for example via hydrogen bonding

It will take more energy for a chemical with a larger size to create the cavity making the chemical more hydrophobic, while interactions of the chemical with water will favour its dissolution making it less hydrophobic. Figure 3 shows chemicals with increasing hydrophobicity with increasing size and a decreasing hydrophobicity by the presence of polar groups (amino or hydroxy).

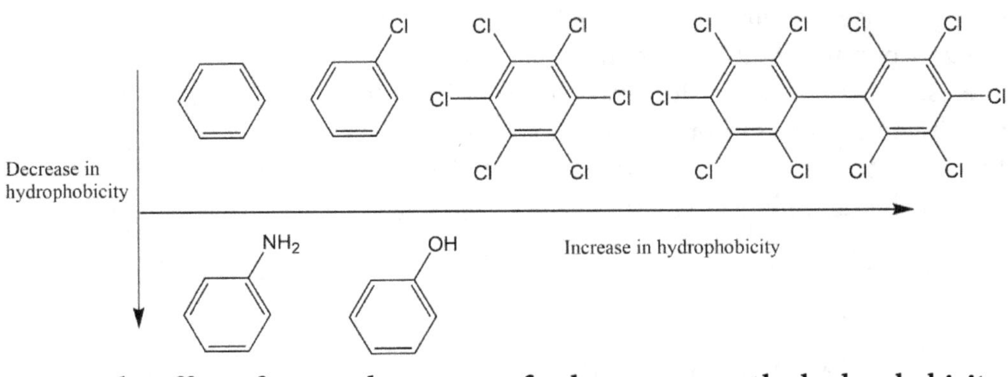

Figure 3. The effect of size and presence of polar groups on the hydrophobicity of chemicals. Increasing molecular size increases hydrophobicity; the introduction of polar groups leads to a decrease in hydrophobicity.

Most hydrophobic chemicals are non-polar organic micro pollutants. Well-known examples are the chlorinated hydrocarbons, such as polychlorinated biphenyls (PCBs) and polycyclic aromatic hydrocarbons (PAHs). Water solubility of these chemicals in general is rather low (in the order of a few ng/L up to a few mg/L).

The hydrophobic nature mainly determines the distribution of these chemicals in water and sediment or soil and their uptake across cell membranes. Additional Cl- or Br-atoms in a chemical, as well as additional (CH)x units, increase the molecular size, and thus a chemical's hydrophobicity. The increased molecular volume requires a larger cavity to dissolve the chemical in water, while they only interact with water molecules via VanderWaals interactions.

Polar groups, such as the -OH and -NH units on the aromatic chemicals in Figure 3, can form hydrogen-bonds with water, and therefore substantially reduce the hydrophobicity of organic chemicals. The hydrogen bonding of hydroxy-substituents works in two ways: The oxygen of -OH bridges to the H-atoms of water molecules, while the hydrogen of –OH can form bridges to the O-atoms of water molecules. Nearly all molecular units consisting of some kind of (carbon-oxygen) combination reduce the hydrophobicity of organic contaminants, because even though they increase the molecular volume they interact via hydrogen bonds (H-bonds) with surrounding water molecules. Additional polar groups in a chemical typically decrease a chemical's hydrophobicity.

Octanol-water Partition Coefficient:

A simple measure of the hydrophobicity of chemicals, originating from pharmacology, is the octanol-water partition coefficient, abbreviated as Kow (and sometimes also called Pow or Poct): this is the ratio of concentrations of a chemical in n-octanol and in water, after establishment of an equilibrium between the two phases (Figure 4). The -OH group in n-octanol does allow for some hydrogen bonding between octanol-molecules in solution, and between octanol and dissolved molecules. However, the relatively long alkyl chain only interacts through VanderWaals interactions, and therefore the interaction strength between octanol-molecules is much smaller than that between water-molecules, and it is energetically much less costly to create a cavity to dissolve any molecule.

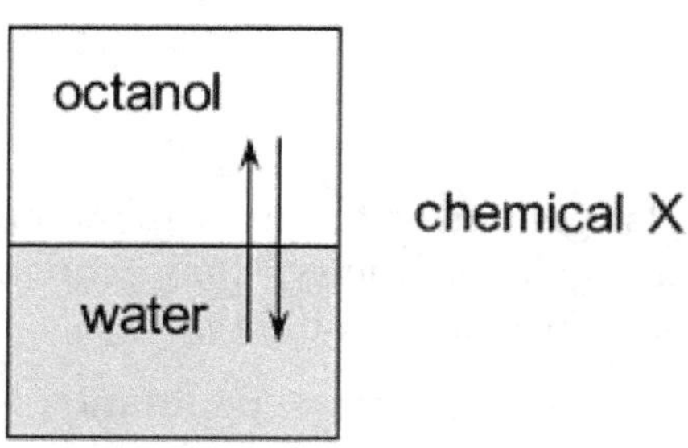

$K_{ow} = C(x)_{octanol} / C(x)_{water}$

Example:
$C(x)_{octanol} = 2.0\ 10^5$ ug/L
$C(x)_{water} = 2.0$ ug/L

$K_{ow} = 10^5$
$\log K_{ow} = 5.0$

Figure 4. Distribution of chemical X between octanol and water and an example of a chemical with log Kow of 5.0.

Experimentally determined Kow values were used in pharmacological research to predict the uptake and biological activity of pharmaceuticals. Octanol was selected because it appears to closely mimic the nonionic molecular properties of most tissue components, particularly phospholipids in membranes. Since the beginning of the 1970s, Kow values have also been used in environmental toxicology to predict the hazard and environmental fate of organic micro pollutants. Octanol may partially also mimic the nonionic molecular properties of most organic matter phases that sorb neutral organic chemicals in the biotic and abiotic environment.

Not unexpectedly, water solubility is negatively correlated with octanol-water partition coefficients.

In practice, three methods can be used to determine or estimate the K_{ow}:

Equilibration Methods

In the shake-flask method (Leo et al., 1971) and the 'slow-stirring' method (de Bruijn et al., 1989), the distribution of a chemical between octanol and water is determined experimentally. For highly lipophilic chemicals ($\log K_{ow} > 5$-6), the extremely low water solubility, however, hampers a reliable analytical determination of concentrations in the water phase. For such chemicals, these experimental methods are not suitable. During the last two decades, the use of generator columns has allowed for quantification of higher Kow values. Generator columns are columns packed with a sorbing material (e.g. Chromosorb®) onto which an appropriate hydrophobic solvent (e.g. octanol) is coated that contains the compound of interest. In this way, a large interface surface area between the lipophilic and water phases is created, which allows for a rapid establishment of equilibrium. When a large volume of (octanol-saturated) water (typically up to 10 litres) is passed slowly through the column, an equilibrium distribution of the compound is established between the octanol and the water. The water leaving the column is passed over a solid sorbent cartridge to concentrate the compound and allow for a quantification of the aqueous concentration. In this way, it is possible to more reliably determine log Kow values up to 6-7.

Chromatography

K_{ow} values may also be derived from the retention time in a chromatographic system (Eadsforth, 1986). The use of reversed-phase High Performance Liquid Chromatography (HPLC), thin-layer chromatography or gas chromatography results in a capacity factor (relative retention time; retention of the compound relative to a non-retained chemical species), which may be used to predict the chemical distribution over octanol and water. HPLC systems have shown most successful, because they consist of stationary and mobile phases that are liquid. As a consequence, the nature of the phases can be most closely arranged to resemble the octanol-water system. Of course, this requires calibration of the capacity factors by applying the chromatographic method to a number of chemicals with well-known Kow values. Chromatographic methods may reliably be applied for estimations of log K_{ow} values in the range of 2-8. For more lipophilic chemicals, also these methods will fail to reliably predict Kow values (Schwarzenbach et al., 2003).

Calculation

Kow values may also be calculated or predicted from parameters describing the chemical structure of a chemical. Several software programs are commercially available for this purpose, such as KOWWIN program of the US-EPA. These programs make use of the so-called fragment method (Leo, 1993; Rekker and Kort, 1979). This method takes into account the contribution to Kow of different chemical groups or atoms in a molecule, and in addition corrects for special features such as steric hindrance or other intramolecular interactions (equation 1):

$$\log Kow = \Sigma fn + \Sigma Fp \qquad \text{(eq.1)}$$

In which fn quantifies the contributions of each fragment n in a particular chemical (see e.g. Table 1) and Fp accounts for any special intramolecular interaction p between the fragments.

This fragment approach has been improved during the last decades and is available in the EPISUITE program from the US Environmental Protection Agency. Other programs for the calculation of Kow values are: ChemProp, and ChemAxon from Chemspider.

Table 1. Fragment constants (Kow) for a few fragments. (from the EPISUITE program)

Fragment	Fragment constant (f)a
-CH3 aliphatic carbon	0.5473
Aromatic Carbon	0.2940
-OH hydroxy, aromatic attach	-0.4802
-N aliphatic N, one aromatic attach	-0.9170

Note: the above calculations are given for non-ionized chemicals. The hydrophobicity of ionic chemicals is also highly affected by the degree of ionization (see below).

K_{ow} values can also be retrieved from databases like echemportal or ECHA and others.

Volatility

Volatility of a chemical from the aqueous phase to air (see Figure 5) is expressed via the Henry's law constant (K_H).

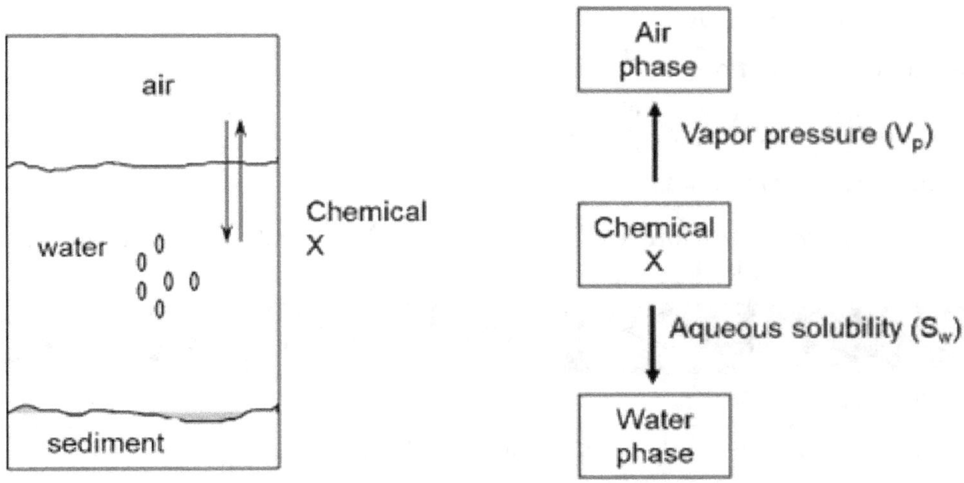

Figure 5. Evaporation of chemical X from water to air.

Henry's law constant (K_H, in Pa·m3/mol) is the chemical distribution between the gas phase and water, as

$$K_H = \frac{P_i}{C_{aq}} \quad \text{(eq.2)}$$

Where in an equilibrated water-gas system:

C_{aq} is the aqueous concentration of the chemical (units in mol/m3), and Pi is the partial pressure of the chemical in air (units in Pascal, Pa), which is the pressure exerted by the chemical in the total gas phase volume (occupied by the mixture of gases the gas-phase above the water solution of the chemical). Note that Pi is a measure of the concentration in the gas phase, but not yet in the same units as the dissolved concentration (discussed below)!

For compounds that are slightly soluble in water, K_H can be estimated from:

$$K_H = \frac{V_p}{S_w} \quad \text{(eq.3)}$$

Where:

K_H: Henry's law constant (Pa·m^3/mol), Vp is the (saturated) vapor pressure (Pa), which is the pressure of the chemical above the pure condensed (liquid) form of the chemical, and Sw is the maximum solubility in water (mol/m3).

The advantage of equation 3 is that both Vp and Sw can be experimentally derived or estimated. The rationale behind equation 3 is that two opposite forces will affect the evaporation of a chemical from water to air:

- the vapor pressure (Vp) of the pure chemical - high vapor pressure means more volatile, and

- solubility in water (Sw) - high solubility means less volatile.

Benzene and ethanol (see Table 2) are good illustrations. Both chemicals have similar vapor pressure, but the Henry's law constant for benzene is much higher because of its much lower solubility in water compared to ethanol; benzene is much more volatile from an aqueous phase.

Table 2. Air-water partition coefficients ($K_{air\text{-}water}$) calculated for five chemicals (ranked by aqueous solubility) by equation 3.

Chemical	Vapor pressure (Pa)	Solubility (mol/m3)	KH (Pa.m3/mol)	Kair-water (L/L, or m3/m3)
Ethanol	7.50·103	1.20·104	6.25·10-1	2.53·10-4
Phenol	5.50·101	8.83·102	6.23·10-2	2.52·10-5
Benzene	1.27·104	2.28·101	5.57·102	2.25·10-1
Pyrene	6.00·10-4	6.53·10-4	9.18·101	3.71·10-4
DDT	2.00·10-5	2.82·10-6	7.08	2.86·10-3

Note: all chemicals at equilibrium have a higher concentration (in e.g. mol/L) in the aqueous phase than in the gas phase. Of these five, benzene is the chemical most prone to leave water, with an equilibrated air concentration about 4 times lower (22.5%) than the dissolved concentration.

Equations 2 and 3 are based on the pressure in the gas phase. Environmental fate is often based on partition coefficients, in this case the air-water partition coefficient ($K_{air\text{-}water}$). These partition coefficients are more or less 'dimensionless', because the concentrations are based on equal volumes (such as L/L), while KH has the unit Pa·m³/mol or something equivalent to the applied units (equation 4).

$$K_{air-water} = \frac{K_H}{RT} \qquad \text{(eq.4)}$$

where:

C_{air} is the concentration in air (in e.g. mol/m3) and Caq is the aqueous concentration (in e.g. mol/m3).

$K_{air\text{-}water}$ can be calculated from KH according to equation 5:

$$K_{air-water} = \frac{K_H}{RT} \qquad \text{(eq.5)}$$

where R is the gas constant (8.314 m3·Pa·K−1·mol−1), and T is the temperature in Kelvin (Kelvin = oCelsius + 273).

This use of "RT" converts this gas phase concentration to a volume based metric, and applies the ideal gas law which relates pressure (P, in Pa) to temperature (T, in

K), volume (V, in m³), and amount of gas molecules (n, in mol), according to the gas constant (R: 8.314 m3·Pa·K-1·mol-1):

P·V = n·R·T (note that the units of both terms will cancel out) (eq.6)

At 25 oCelcius (298 K), the product RT equals 2477 m3·Pa·K-1·mol-1.

Examples of calculated values for $K_{air\text{-}water}$ are presented in Table 2.

The influence of the chemical structure on volatility of a chemical from a solvent fully depends on the cost of creating a cavity in the solvent (interactions between solvent molecules) and the interactions between the chemical and the solvent molecules. For partitioning processes, the gas phase is mostly regarded as an inert compartment without chemical interactions (i.e. gas phase molecules hardly ever touch each other).

The molecules of a strongly dipolar solvent such as water that contain atoms that can interact as hydrogen acceptor (the O in an OH group) and hydrogen donor (the H in an OH group) strongly interact with each other, and it costs much energy to create a cavity. This cost increases strongly with molecular size, for nearly all molecules more than the energy regained by interactions with the surrounding solvent molecules. As a result, for most classes of organic chemicals, affinity with water decreases and volatility out of water into air slightly increases with molecular volume. For chemicals that are not able to re-interact via hydrogen bonding, e.g. alkanes, the overall volatility is much higher than for chemicals that do have specific interactions with water molecules besides van der Waals.

Degree of Ionization

Acids and bases can be present in the neutral (HA and B) or ionized form (A- and BH+, respectively). For acids, the neutral form (HA) is in equilibrium with the anionic form (A-) and for bases the neutral form (B) is in equilibrium with the cationic form (BH+). The degree of ionization depends on the pH and the acid dissociation constant (pKa). Table 3 shows the equations to calculate the fraction ionized for acids and bases and examples of two acids (phenols) are presented in Table 4.

Table 3. Calculation of the fraction ionized for acids and bases.

Acids	Bases
$HA \rightleftharpoons A^- + H^+$	$BH^+ \rightleftharpoons B + H^+$
$fraction\,ionized = \frac{1}{1+10^{(pH-pK_a)}}$	$fraction\,ionized = \frac{1}{1+10^{(pH-pK_a)}}$

$pK_a = -\log K_a$, where K_a is dissociation constant of the acidic form (HA or BH+).

The degree of ionization is thus determined by the pH and the pKa value and more examples for several organic chemicals are presented elsewhere (see Chapter Ionogenic organic compounds).

Table 4. The degree of ionization of two phenolic structures (acids).

<table>
<tr><th>Pentachlorophenol</th><th>Phenol</th></tr>
<tr><td>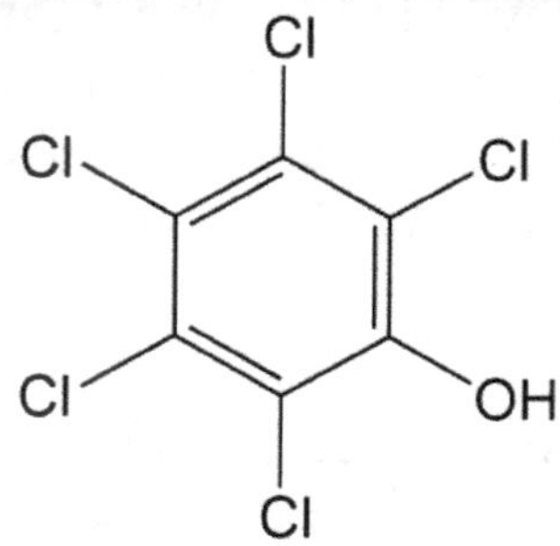
</td><td>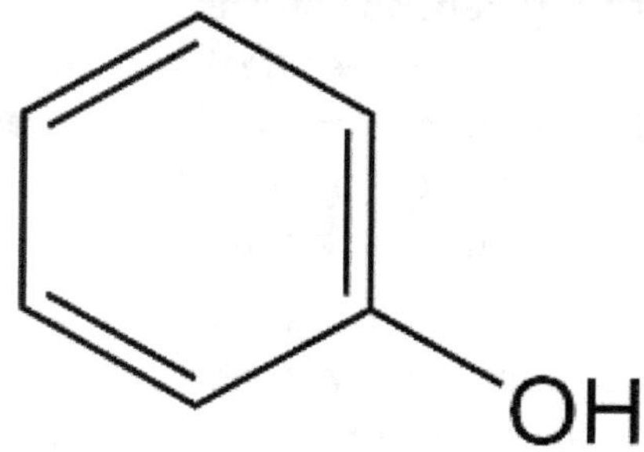
</td></tr>
<tr><td>pKa = 4.60</td><td>pKa = 9.98</td></tr>
<tr><td>% ionized versus pH</td><td>% ionized versus pH</td></tr>
<tr><td>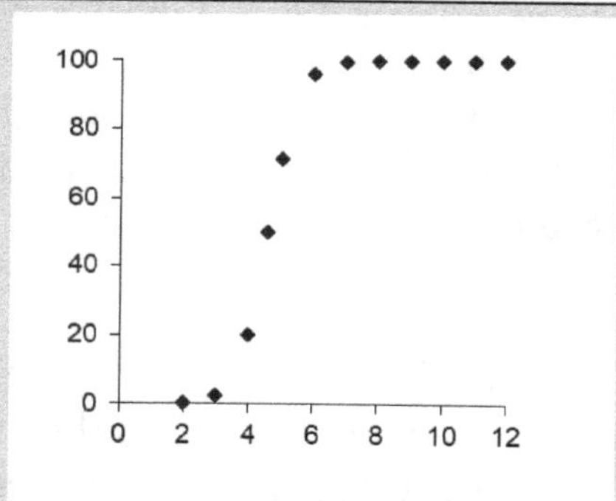
</td><td>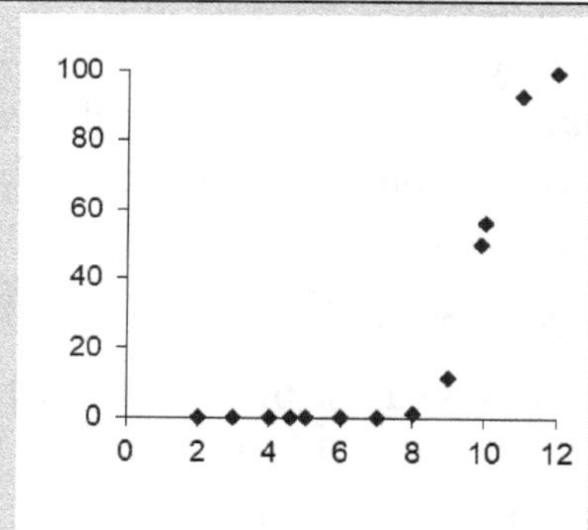
</td></tr>
<tr><td>at pH of 7.0: 99.6 % ionized</td><td>at pH of 7.0: 0.1 % ionized</td></tr>
</table>

Examples for several organic chemicals are presented elsewhere (see section on Ionogenic organic compounds).

The fate of ionic chemicals is very different from that of non-ionic chemicals. The sediment-water sorption coefficient of the anionic species is substantially (>100x) lower than that of the neutral species. If the percentage of ionization is less than ~99 % (at a pH 2 units above the pKa), the sorption of the anion may be neglected (Kd is still dominated by the >1% neutral species) (Schwarzenbach et al., 2003). The reason for the low sorption affinity of the anionic acid form is twofold: anions are much better water soluble, but also most sediment particles (clay, organic matter, silicates) are negatively charged, and electrostatically repulse the similarly charged chemical. In that case the sorption coefficient Kd can be calculated from the sorption coefficient of the non-ionic form and the fraction of the non-ionized form (α):

$$K_d = \alpha K_d\,(\mathit{neutral\ form}) \qquad \text{(eq. 7)}$$

In environments where the pH is such that the neutral acid fraction <1% (when pH >2 units above the pKa), the sorption of the anionic species to soil/sediment may significantly contribute to the overall "distribution coefficient" of both acid species.

For basic environmental chemicals of concern, among which many illicit drugs (e.g. amphetamine, cocaine) and non-illicit drugs (e.g. most anti-depressants, beta-blockers), the protonated forms are positively charged. These organic cations are also much more soluble in water than the neutral form, but at the same time they are electrostatically attracted to the negatively charged sediment surfaces. As a result, the sorption affinity of organic cations to sediment should not be considered negligible relative to the neutral species. The sorption processes, however, may strongly differ for the neutral base species and the cationic base species. Several studies have shown that the sorption affinity of cationic base species to DOM or sediment is even stronger than that of the neutral species.

References

De Bruijn, J., Busser, F., Seinen, W., Hermens, J. (1989). Determination of octanol/water partition coefficients for hydrophobic organic chemicals with the "slow-stirring" method. Environmental Toxicology and Chemistry 8, 499-512.

Eadsforth, C.V. (1986). Application of reverse-phase HPLC for the determination of partition coefficients. Pesticide Science 17, 311-325.

Leo, A., Hansch, C., Elkins, D. (1971). Partition coefficients and their uses. Chemical Reviews 71, 525-616.

Leo, A.J. (1993). Calculating log P(oct) from structures. Chemical Reviews 93, 1281-1306.

Rekker, R.F., de Kort H.M. (1979). The hydrophobic fragmental constant; an extension to a 1000 data point set. Eur.J. Med. Chem. - Chim. Ther. 14:479-488.

Schwarzenbach RP, Gschwend PM, Imboden DM (Eds.) 2003. Environmental organic chemistry. Wiley, New York, NY, USA.

Further Reading:

Mackay, D., Boethling, R.S. (Eds.) 2000. Handbook of property estimation methods for chemicals: environmental health and sciences. CRC Press.

van Leeuwen, C.J., Vermeire, T.G. (Eds.) 2007. Risk assessment of chemicals: An introduction. Springer, Dordrecht, The Netherlands.

3.4.2. Sorption

Introduction

Sorption processes have a major influence on the fate of chemicals in the environment (Box 1). In general, sorption is defined as the binding of a dissolved or gaseous chemical (the sorbate) to a solid phase (the sorbent) and this may involve different processes, including:

- binding of dissolved chemicals from water to sediments and soils and
- binding of gaseous phase chemicals from air to soils, plants, and trees.

Information about sorption is relevant because of a number of reasons:

- sorption controls the actual fate and thereby the risk of (many) organic and inorganic contaminants in the environment,
- sorbed chemicals cannot evaporate, are not available for (photo) chemical or microbial breakdown, are not as easily transported as dissolved/vapor phase chemicals, and are not available for uptake by organisms,
- sorption also plays an important role in toxicity tests, affecting exposure concentrations.

Box 1.

The Biesbosch is a wetland area in the Netherlands, an area in between the Rivers Rhine and Meuse and estuaries that are connected to the North Sea. The water flow is relatively low and as a consequence there is a strong sedimentation of particles from the water to the sediment. Chemicals present in the water strongly sorb to these particles, which in the past were polluted with hydrophobic organic contaminants such as dioxins and PCBs. The concentrations of these organic compounds in sediment are still relatively high because they are highly persistent. The reason for this persistence of these compounds is that these sorbed compounds are not easily available for degradation by bacteria. Also, the concentrations in organisms that live close to or in the sediment are high. These concentrations are so high that fishing on eel, for example, is not allowed in the area. This example shows the importance of sorption processes on fate, but also on effects in the environment.

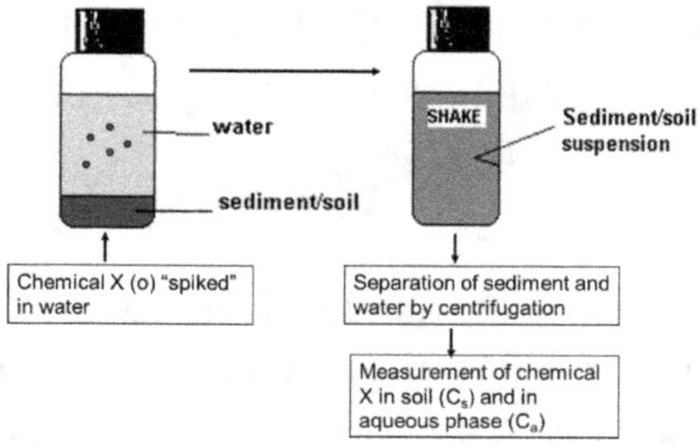

Figure 1. Measurement of sorption coefficients.

Measurement of sorption is a simple procedure. A chemical X is spiked (added) to the aqueous phase in the presence of a certain amount of the solid phase (sediment or soil). The chemical sorbs to the solid phase and when the system is in equilibrium, the concentrations in the sediment (C_s) and in the aqueous phase (Ca) are measured. The solid phase is collected via centrifugation or filtration.

The sorption coefficient K_p (equation 1 and box 2) gives information about the degree of sorption of a chemical to sediment and is defined as:

$$K_p = \frac{C_s}{C_a} \tag{1}$$

Box 2:

The concentration of a chemical X in sediment (Cs) is 30 mg/kg and the concentration in the aqueous phase (Ca) is 0.1 mg/L.

The sorption coefficient Kp = Cs / Ca = 30 mg/kg / 0.1 mg/L = 300 L/kg

Note the units of a sorption coefficient: L/kg

In the environmental risk assessment of chemicals, it is very useful to understand the fraction of the total amount of chemical (Atotal) in a system that is sorbed (*f*sorbed) or dissolved (*f*dissolved) (e.g. due to an accidental spill in a river):

*f*dissolved = Adissolved / Atotal , and thus *f*sorbed = 1 - *f*dissolved

This is related to the sorption coefficients of X and the volume of the solvent and the volume of the sorbent material. The equation derived for calculating *f*dissolved is based on the mass balance of chemical A, which relates the concentration of X (C) to the amount of X (A) in each volume (V):

C = A / V, and thus A = C·V

which for a system of water and sediment (air not included for simplification) relates to:

Atotal = Adissolved + Asorbed = Cwater · Vwater + Csediment · Vsediment = Cwater · Vwater + (Kp·Cwater)·Vsediment

*f*dissolved = Adissolved / Atotal = Cwater · Vwater / (Cwater ·Vwater + Kp ·Cwater·Vsediment)

This way of separating out Csediment from the equation using Kp can result, after rearranging (by dividing both parts of the ratio by Cwater · Vwater) to the following simplified equation:

*f*dissolved = 1 / (1 + Kp·(Vsediment / Vwater))

In this equation, 'sediment' can be replaced by any sorbent, as long as the appropriate sorption coefficient is used.

Let's try to calculate with chemical X from above, in a wet sediment, where 1L wet sediment contains ~80% water and 20% dry solids. The dissolved fraction of X with Kp = 300 kg/L, is only 0.013 in this example. Thus, with 1.3% of X actually dissolved, this indicates that 98.7% of X is sorbed to the sediment.

Sorption Processes

There are two major sorption processes (see Figure 2):

- **Absorption** - Partitioning ("dissolution") of a chemical into a 3-D sorbent matrix. The concentration in the sorbing phase is homogeneous.
- **Adsorption** - Binding of a chemical to a 2-D sorbent surface. Because the number of sorption sites on a surface is limited, sorption levels off at high concentrations in the aqueous phase.

A sorption isotherm gives the relation between the concentration in a sorbent (sediment) and the concentration in the aqueous phase and the isotherm is important in identifying a sorption process.

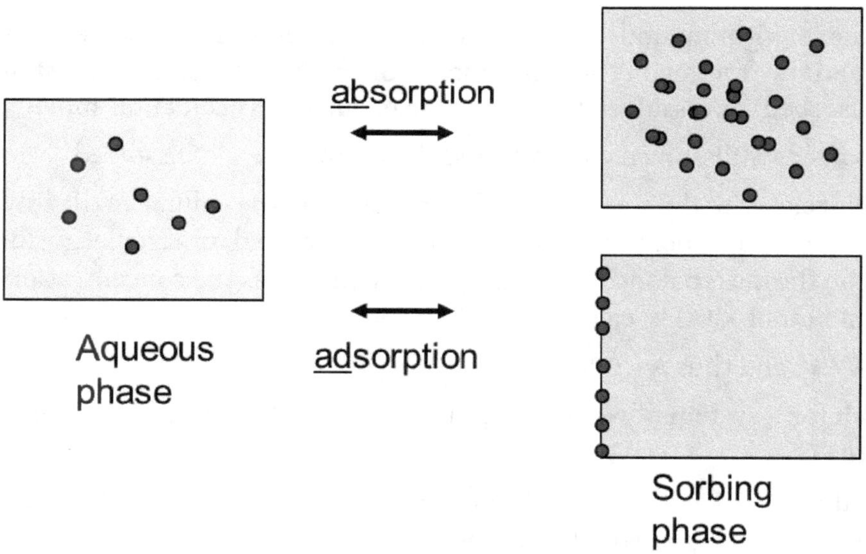

Figure 2. Two sorption processes: absorption and adsorption.

Absorption of a chemical is similar to its partitioning between two phases and comparable to its partitioning between two solvents. Distribution of a chemical between octanol and water is a well-known example of a partitioning process (see Section 3.4.1 on Relevant chemical properties for more detailed information on octanol-water partitioning). The isotherm for an absorption process is linear (Figure 3A) and the slope of the y-x plot is the sorption coefficient Kp.

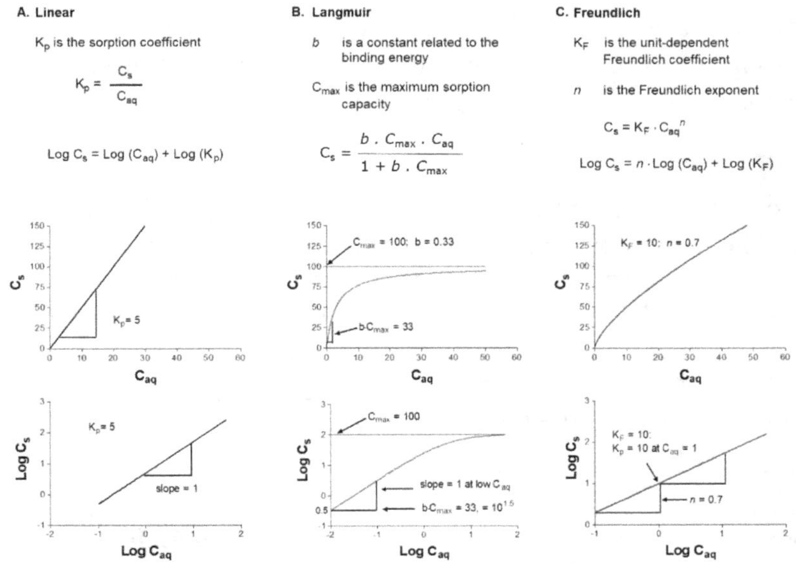

Figure 3. Sorption isotherms for absorption: linear model (3A), and for adsorption: Langmuir model (3B) or Freundlich model (3C).

In an adsorption process, where the sorbing phase is a surface with a limited number of sorption sites, the sorption isotherm is non-linear and may reach a maximum concentration that is adsorbed when all sites are occupied. A mechanistic model for adsorption is the Langmuir model. This model describes adsorption of molecules to homogeneous surfaces with equal adsorption energies, represented by the adsorption site energy term (b) and a limited number of sorption sites (Cmax) that can become saturated (Figure 3B). The Langmuir adsorption coefficient (Kad) is equal to the product (b · Cmax) at relatively low aqueous concentrations, where the product (b · Caq) << 1 (note that the denominator term will then be ~1). Indeed, the isotherm curve on a double log scale plot shows a slope of 1 at such low concentrations, indicating linearity.

Another mathematical approach to describe non-linear sorption is the Freundlich isotherm (Figure 3C), where KF is the Freundlich sorption constant and n is the Freundlich exponent describing the sorption process non-linearity. Using logarithmic values for aqueous and sorbed concentrations, the Freundlich isotherm can be rewritten as:

$$\text{Log } C_s = n \cdot \log C_{aq} + \log K_F \qquad \text{(eq. 2)}$$

This conveniently yields a linear relationship (just as y = a·x + b) between log Cs and log Caq, with a slope equal to n and the abscissa (crossing point with the Y-axis) equal to log KF. This allows for easy fitting of linear trend lines through experimental data sets. When n = 1, the isotherm is linear, and equals the one for absorption. In case of saturation of the sorption sites on the solid phase, 1/n will be smaller than 1. The Freundlich isotherm can, however, also yield a 1/n value > 1; this may occur for example if the chemical that is sorbed itself forms a layer that serves as a new sorbing phase and examples are described for surfactants.

Sorption Phases

Soils and sediments may show large variations in composition and particle size distribution. The major components of soils and sediments are:

Sand	63 – 2 mm
Silt	2 – 63 µm
Clay	<2 µm
Organic matter	includes e.g. detritus, humic acids, especially associated with the clay and silt fractions
$CaCO_3$	

Figure 4 gives a schematic picture of a sediment or soil particle. In addition to the presence of clay minerals and (soil or sediment) organic matter (SOM), sediment and soil may contain soot particles (a combustion residue).

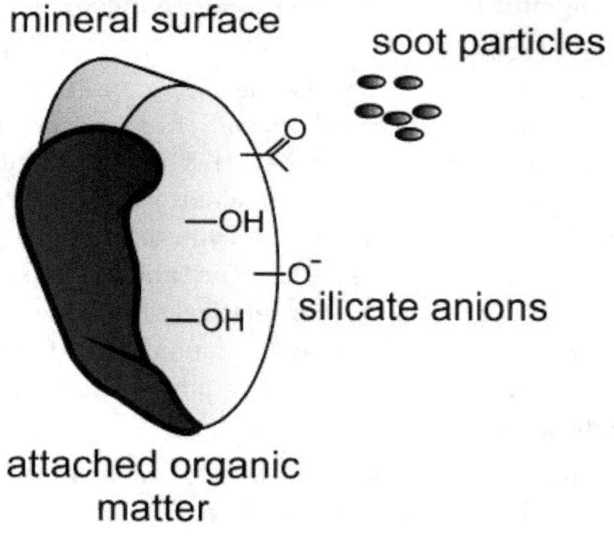

Figure 4. Structure of a soil or sediment particle showing the major components: organic matter and clay mineral and soot. Modified from Schwarzenbach et al. (2003) by Steven Droge.

Organic matter is formed upon decomposition of plant material and dead animal or microbial tissues. Upon decomposition of plant material, the first organic groups to be released are phenolic acids, some of which have a high affinity for complexation of metals. One example is salicylic acid (o-hydroxybenzoic acid), which occurs in high concentrations in leaves of willows, poplar and other deciduous trees. Further decomposition of plant material may result in the formation of humic acids, fulvic acids and humin. Humic and fulvic acids contain a series of functional groups, such as carboxyl- (COOH), carbonyl- (=C=O), phenolic hydroxyl- (-OH), methoxy- (-OCH3), amino- (-NH2), imino (=NH) and sulfhydryl (-SH) groups (see for more details the section on Soil).

Hydrophobic organic chemicals mainly sorb to organic matter. Because organic matter has the characteristics of a solvent, the sorption is clearly an absorption process and the sorption isotherm is linear. Because binding is mainly to organic matter, the sorption coefficient (Kp) depends on the fraction of organic matter (fom) or the fraction of organic carbon (foc) present in the soil or sediment. Please note that as a rule of thumb, organic matter contains 58% organic carbon (foc = 0.58·fom). Figure 5A shows the increase in sorption coefficient with increasing fraction organic carbon in soils and sediments. In order to arrive at a more intrinsic parameter, sorption coefficients are often normalized to the fraction organic matter (Kom) or organic carbon (Koc). These Koc or Kom values are less dependent of the sediment or soil type (Figure 5B).

$$K_{om} = \frac{K_p}{f_{om}} \tag{3}$$

$$K_{oc} = \frac{K_p}{f_{oc}} \tag{4}$$

Figure 5. The relationship between the sorption coefficient (Kp) (left) and the organic carbon normalized sorption coefficient (Koc) (right) and the fraction organic carbon (foc). Data from Means et al. (1980). Drawn by Wilma Ijzerman.

Hydrophobic chemicals can have a very high affinity to soot particles relative to the affinity to SOM. If a sediment contains soot, Kp values are often higher than predicted based on the fraction organic carbon in the organic matter (Jonker and Koelmans, 2002).

References

Schwarzenbach, R.P., Gschwend, P.M., Imboden, D.M. (2003). Environmental Organic Chemistry. Wiley, New York, NY, USA.

Means, J.C., Wood, S.G., Hassett, J.J., Banwart, W.L. (1980). Sorption of polynuclear aromatic-hydrocarbons by sediments and soils. Environmental Science and Technology 14, 1524-1528.

Jonker, M.T.O., Koelmans, A.A. (2002). Sorption of polycyclic aromatic hydrocarbons and polychlorinated biphenyls to soot and soot-like materials in the aqueous environment mechanistic considerations. Environmental Science and Technology 36, 3725-3734.

Suggested Reading

van Leeuwen, C.J., Vermeire, T.G. (Eds.) (2007). Risk Assessment of Chemicals: An Introduction. Springer, Dordrecht, The Netherlands. Chapter 3 and 9.

Schwarzenbach, R.P., Gschwend, P.M., Imboden, D.M. (2003). Environmental Organic Chemistry. Wiley, New York, NY, USA. chapters 9, 11. Detailed information about sorption processes and sorption mechanisms.

3.4.3. QSPRs

Introduction

Risk assessment needs input data for fate and effect parameters. These data are not available for many of the existing chemicals and predictions via estimation models will provide a good alternative to actual testing. Examples of estimation models are Quantitative Structure-Property Relationships (QSPRs) and Quantitative Structure-Activity Relationships (QSARs). The term “activity” is often used in relation to models for toxicity, while “property” usually refers to physical-chemical properties or fate parameters.

In a QSAR or QSPR, a certain environmental parameter is related to a physical-chemical or structural property, or a combination of properties.

The elements in a QSPR or QSAR are shown in Figure 1 and include:

- The parameter for which the estimation model has been developed: the Y-variable (upper right),
- the properties of the chemical or chemical parameter: the X-variable (upper left),
- the model itself (center), and
- the prediction of a fate or effect parameter from the chemical properties (bottom).

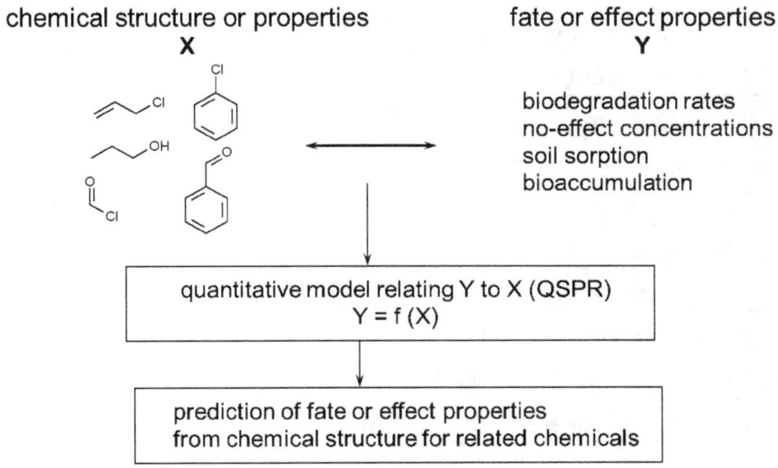

Figure 1. The principle of a QSPR or QSAR. See text for explanation.

The Y-variable

Estimation models have been developed for many endpoints such as sorption to sediment, humic acids, lipids and proteins, chemical degradation, biodegradation, bioconcentration and ecotoxic effects.

The X-variable

An overview of the chemical parameters (the X-variable) used in estimation models is given in Table 1. Chemical properties are divided in three categories: (i) parameters related to hydrophobicity, (ii) parameters related to charge and charge distribution in a molecule and (iii) parameters related to the size or volume of a molecule. Hydrophobicity is discussed in more detail in the section on Relevant chemical properties.

Other QSPR approaches use large number of parameters derived from chemical graphs. The CODESSA Pro software, for example, generates molecular (494) and fragment (944) descriptors, classified as (i) constitutional, (ii) topological, (iii) geometrical, (iv) charge related, and (v) quantum chemical (Katritzky et al. 2009). Some models are based on structural fragment in a molecule. The polyparameter linear free energy relationships (pp-LFER) use parameters that represent interactions between molecules (see under pp-LFER).

Table 1. Examples of parameters related to hydrophobicity and electronic and steric parameters (the X variable).

Hydrophobic parameters
Aqueous solubility
Octanol-water partition coefficient (K_{ow})

Hydrophobic fragment constant π
Electronic parameters
Atomic charges (q)
Dipole moment
Hydrogen bond acidity (H bond-donating)
Hydrogen bond basicity (H bond-accepting)
Hammett constant σ
Steric parameters
Total Surface Area (TSA)
Total Molecular Volume (TMV)
Taft constant for steric effects (Es)

The Model

Most models are based on correlations between Y and X. Such a relationship is derived for a "training set" that consists of a limited number of carefully selected chemicals. The validity of such a model should be tested by applying it to a "validation set", i.e. a set of compounds for which experimental data can be compared with the predictions. Different techniques can be used to develop an empirical model, such as:

- Graphical presentations,
- linear or non-linear equations between Y and X,
- linear or non-linear equations based on different properties (Y versus X_1, X_2, etc.),
- multivariate techniques such as Principal Component Analysis (PCA) and Partial Least Square Analysis (PLS).

Linear equations take the form:

$$Y(i) = a_1X_1(i) + a_2X_2(i) + a_3X_3(i) + \ldots + b \qquad (1)$$

where Y(i) is the value of the dependent parameter of chemical i (for example sorption coefficients); X1-X3(i) are values for the independent parameters (the chemical properties) of chemical i; a1-a3 are regression coefficients (usually 95% confidence limits are given); b is the intercept of the linear equation. The quality of the equation is presented via the correlation coefficient (r) and the standard error of estimate (s). The closer r is to 1.0, the better the fit of the relationship is. More information about the statistical quality of models can be found under "limitation of QSPR".

The classical approach in QSAR and QSPR studies is the Hansch approach that was develop in the 1960s. The Hansch equation (Hansch et al., 1963) describes the influence of substituents on the biological activity in a series of parent compounds with a certain substituent (equation 2). Substituents are for example a certain atom or chemical group (Cl, F, B,. OH, NH2) attached to a parent aromatic ring structure.

$$\log 1/C = c\,\pi + c'\,\sigma + c''\,E_s + c''' \quad (2)$$

In which:

- C is the molar concentration of a chemical with a particular effect,
- π is a substituent constant for hydrophobic effects,
- σ is a substituent constant for electronic effects, and
- Es is a substituent constant for steric effects.
- c are constants that are obtained by fitting experimental data

For example, the hydrophobic substituent constant is based on K_{ow} and is defined as is defined as:

$$\pi\,(X) = \log K_{ow}\,(RX) - \log K_{ow}\,(RH) \quad (3)$$

where RX and RH are the substituted and unsubstituted parent compound, respectively.

The Hammett and Taft constants are derived in a similar way.

Multivariate techniques may be very useful to develop structure-activity relationships, in particular in cases where a large number of chemical parameters is involved. Principal Component Analysis (PCA) can be applied to reduce the number of variables into a few principal components. The next step is to find a relationship between Y and X via, for example, Partial Least Square (PLS) analysis. The advantage of PCA and PLS is that it can deal with a large number of chemical descriptors and that is can also cope with collinear (correlated) properties. More information on these multivariate techniques and examples in the field of environmental science are given by Eriksson et al. (1995).

Poly-parameter Linear Free Energy Relationship (pp-LFER)

The pp-LFER approach has a strong mechanistic basis because it includes the different types of interactions between molecules (Goss and Schwarzenbach, 2001). For example, the sorption coefficient of a chemical from an aqueous phase to soil or to phospholipids (the sorbent) depends on the interaction of a chemical with water and the interaction with the sorbent phase. One of the driving forces behind sorption is the hydrophobicity. Hydrophobicity means fear (phobia) for water (hydro). A hydrophobic chemical prefers to “escape from the aqueous phase” or in other words “it does not like to dissolve in water”. Water molecules are tightly bound to each other via hydrogen bonds. For a chemical to dissolve, a cavity should be formed in the aqueous phase (Figure 2) and this will cost energy. More hydrophobic compounds will often have a stronger sorption (see more information in the section on Relevant chemical properties).

Hydrophobicity mainly depends on two molecular properties:

- Molecular size
- Polarity / ability to interact with water molecules, for example via hydrogen bonding

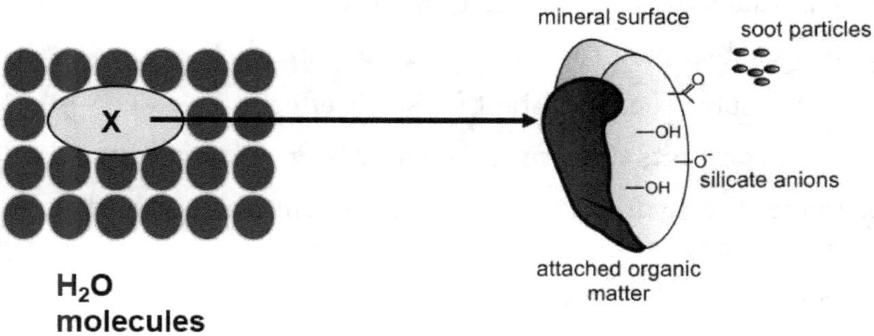

Figure 2. The formation of a cavity in water for chemical X and the interaction with another phase (here, a soil particle).

In the interaction with the sorbent (soil, membrane lipids, storage lipids, humic acids), major interactions are van der Waals interactions and hydrogen bonding (Table 2). Van der Waals interactions are attractive and occur between all kind of molecules and the strength depends on the contact area. Therefore, the strength of van der Waals interactions are related to the size of a molecule. A hydrogen bond is an electrostatic attraction between a hydrogen (H) and another electronegative atom bearing a lone pair of electrons. The hydrogen atom is usually covalently bound to a more electronegative atom (N, O, F). Table 2 lists the interactions with examples of chemical structures.

A pp-LFER is a linear equation developed to model partition or sorption coefficients (K) using parameters that represent the interactions (Abraham, 1993). The model equation is based on five descriptors:

$$logK = c + e \bullet E + s \bullet S + a \bullet A + b \bullet B + v \bullet V \tag{2}$$

With:

E	excess molar refraction
S	dipolarity/polarizability parameter
A	solute H-bond acidity (H-bond donor)
B	solute H-bond basicity (H-bond acceptor)
V	molar volume

The partition or sorption coefficient K may be expressed as the sum of five interaction terms, with the uppercase parameters describing compound specific properties. E depends on the valence electronic structure, S represents polarity and polarizability, A is the hydrogen bond (HB) donor strength (HB acidity), B the HB acceptor strength (HB basicity), V is the so-called characteristic volume related to the molecule size, and c is a constant. The lower-case parameters express the corresponding properties of the respective two-phase system, and can thus be taken as the relative importance of the compound properties for the particular partitioning or sorption process. In this introductory section, we only focus on the volume factor (V) and the two hydrogen bond parameters (A and B).

Numerous pp-LFERs have been developed for all kinds of environmental processes and an overview is given by Endo and Goss (2014).

Table 2. Types of interactions between molecules and the phase to which they sorb with examples of chemicals (Goss and Schwarzenbach, 2003).

Compound[a)]	Interactions	Examples
Apolar	only van der Waals	alkanes, chlorobenzenes, PCBs
Monopolar	van der Waals + H-acceptor (e-donor)	alkenes, alkynes, alkylaromatic compounds ethers, ketones, esters, aldehydes
Monopolar	van der Waals + H-donor (e-acceptor)	CHCl3, CH2Cl2
Bipolar	van der Waals + H-donor + H-acceptor	R–NH2, R2–NH, R–COOH, R–OH

[a)] **Apolar:** no polar group present; mono/dipolar: one or two polar groups present in a molecule

Examples of QSPR for Bioconcentration to Fish

K_{ow} based model

Predictive models for bioconcentration have a long history. The octanol-water partition coefficient (K_{OW}) is a good measure for hydrophobicity and bioconcentration factors (BCF's) are often correlated to K_{ow} (see more information in section on Bioaccumulation). The success of these K_{OW} based models was explained by the resemblance of partitioning in octanol and bulk lipid in the organisms, at least for neutral hydrophobic compounds. A well-known example of a linear QSAR model for the log BCF (Y variable) based on the log KOW (X variable) (Veith et al., 1979):

$$\log BCF = 0.85 \log K_{ow} - 0.70 \quad (5)$$

Figure 3 gives a classical example of such a correlation for BCF to guppy of a series of chlorinated benzenes and polychlorinated biphenyls. When lipophilic chemicals are metabolised, the relation shown in Figure 3 is no longer valid and BCF will be lower than predicted based on KOW. Another deviation of this BCF-Kow relation can be found for highly lipophilic chemicals with log Kow>7. For such chemicals, BCF often decrease again with increasing Kow (see Figure 3).

The apparent BCF curve with Kow as the X variable tends to follow a nonlinear curve with an optimum at log Kow 7-8. This phenomenon may be explained from molecular size: molecules of chemicals like decachlorobiphenyl may be so large that they have difficulties in passing membranes. A more likely explanation, however, is that for highly lipophilic chemicals aqueous concentrations may be overestimated. It is not easy to separate chemicals bound to particles from the aqueous phase (see box 1 in the section on Sorption) and this may lead to measured concentrations that are higher than the bioavailable (freely dissolved) concentration (Jonker and van der Heijden 2007; Kraaij et al. 2003). For example, at a dissolved organic carbon (DOC) concentration of 1 mg-DOC/L, a chemical with a log Koc of 7 will be 90% bound to particles, and this bound fraction is not part of the dissolved concentration that equilibrates with the (fish) tissue. This shows that these models are also interesting because they may show trends in the data that may lead to a better understanding of processes.

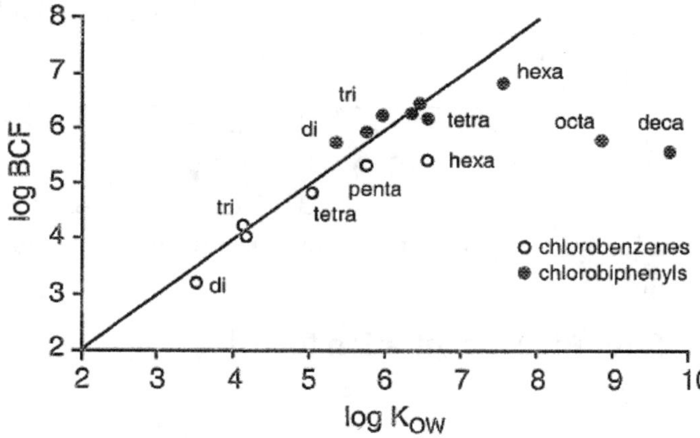

Figure 3. The relationship between bioconcentration factors in guppy and the octanol-water partition coefficients with data from (Bruggeman et al., 1984; Könemann and Leeuwen, 1980).

Examples of QSPR for sorption to lipids

K_{ow} based models are successful because octanol probably has similar properties than fish lipids. There are several types of lipids and membrane lipids have different properties and structure than for example storage lipids (see Figure 4, and more details in the section on Biota). More refined BCF models include separation of storage and membrane lipids and also proteins as separate sorptive phases (Armitage et al. 2013). pp-LFER is a very suitable approach to model these sorption or partitioning processes and results for two large data sets are presented in Table 3. The coefficients e, s, b and v are rather similar. The only parameter that is different in these two models is coefficient a, which represents the contribution of hydrogen bond (HB) donating properties (A) of chemicals in the data set. This effect makes sense because the phosphate group in the phospholipid structure has strong HB accepting properties. This example shows the strength of the pp-LFER approach because it closely represents the mechanism of interactions.

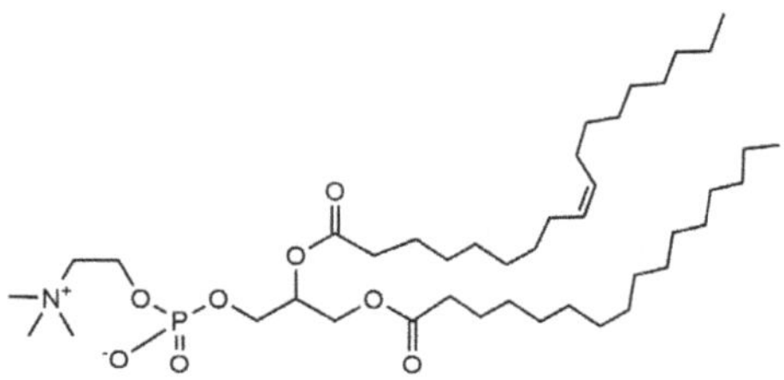

1-palmitoyl-2-oleoylphosphatidylcholine (POPC), a key phospholipid in cell membranes

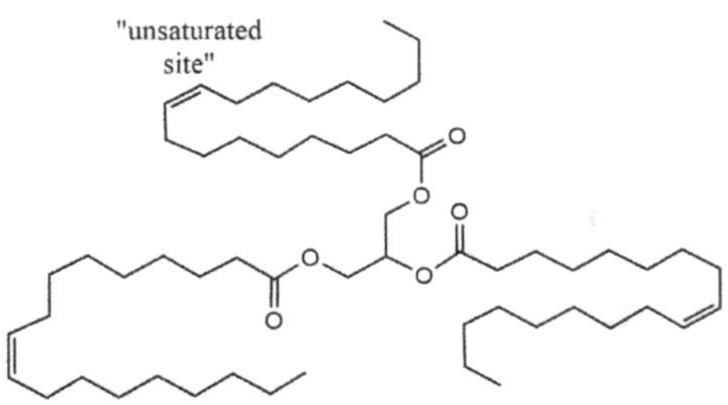

Triglyceride, storage fat

Figure 4. Structure of a phospholipid and a triglyceride. Note the similar glycerol part in both lipids.

Table 3. LFERs for storage lipid-water partition coefficients (K_{SL-W}) and membrane lipid-water partition coefficients (K_{ML-W} (liposome)). Listed are the parameters (and standard error), the number of compounds with which the LFER was calibrated (n), the correlation coefficient (r^2), and the standard error of estimate (SE). log K = c + e*E* + s*S* + a*A* + b*B* + v*V*.

Parameter	c	e	s	a	b	v	n	r2	SE
K_{SL-W}	-0.07 (0.07)	0.70 (0.06)	-1.08 (0.08)	-1.72 (0.13)	-4.14 (0.09)	4.11 (0.06)	247	0.997	0.29
From (Geisler et al. 2012)									
KML-W (liposome)	0.26 (0.08)	0.85 (0.05)	-0.75 (0.08)	0.29 (0.09)	-3.84 (0.10)	3.35 (0.09)	131	0.979	0.28
From (Endo et al. 2011)									

K_{SL-W}: storage lipid partition coefficients are mean values for different types of oil. Raw data and pp-LFER (for 37 °C) reported in (Geisler et al. 2012).

K_{ML-W} (liposome): data from liposomes made up of phosphatidylcholine (PC) or PC mixed with other membrane lipids. Raw data (20-40 °C) and pp-LFER reported in (Endo et al. 2011).

Examples of QSPR for Sorption to Soil

Numerous QSPRs are available for soil sorption (see section on Sorption). Also the organic carbon normalized sorption coefficient (K_{oc}) is linearly related to the octanol-water partition coefficient (see Figure 5).

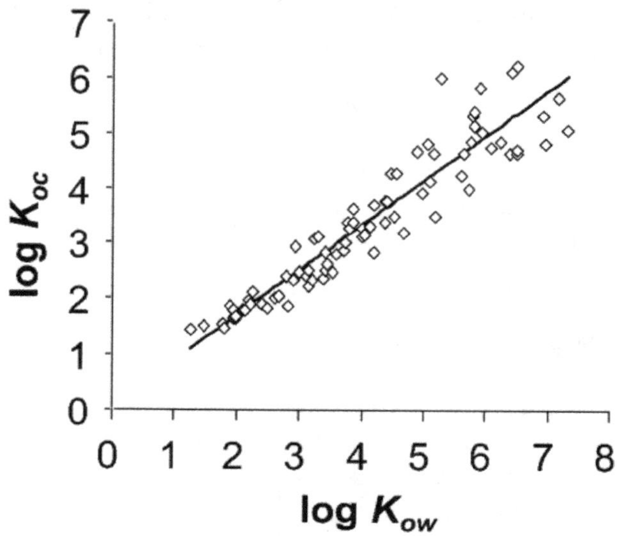

Figure 5. Correlation between the organic carbon normalized sorption coefficient to soil (Koc) and the octanol-water partition coefficient (Kow) for data from (Sabljic et al. 1995).

The model in Figure 5 is only valid for neutral, non-polar hydrophobic organic chemicals such as chlorinated aromatic compounds, polycyclic aromatic hydrocarbons (PAHs), polychlorinated biphenyl (PCBs) and chlorinated insecticides or, in general, compounds that only contain carbon, hydrogen and halogen atoms. It does not apply to polar and ionized organic compounds nor to metals. For polar chemicals, also other interactions may influence sorption and a pp-LFER approach would also be useful.

The sorption of ionic chemicals is more complex. For the sorption of cationic organic compounds, clay minerals can be an equally important sorption phase as organic matter because of their negative surface charge and large surface area. The sorption of organic cations is mainly an adsorption process that reaches a maximum at the cation exchange capacity (CEC) of a particle (see section on Soil). Also models for the prediction of sorption of cationic compounds are more complicated and first attempts have been made recently (Droge and Goss, 2013). Major sorption mechanism for anionic chemicals is sorption into organic matter. The sorption coefficient of anionic chemicals is substantially lower than for the neutral form of the chemical, roughly a factor 10-100 for KOC (Tülp et al. 2009). In case of weakly dissociating chemicals such as carboxylic acids, the sorption coefficient can often be estimated from the sorption coefficient of the non-ionic form and the fraction of the chemical that is present in the non-ionized form (see section on Relevant chemical properties).

Reliability and limitations of QSPR

Predictive models have limitations and it is important to know these limitations. There is not one single model that can predict a parameter for all chemicals. Each model will have a domain of applicability and it is important to apply a model only to a chemical within that domain. Therefore, guidance has to be defined on how to select a specific model. It is also important to realize that in many computer programs (such as fate modeling programs), estimates and predictions are implicitly incorporated in these progams.

Another aspect is the reliability of the prediction. The model itself can show a good fit (high r^2) for the training set (the chemicals used to develop the model), but the actual reliability should be tested with a separate set of chemicals (the validation set) and a number of statistical procedures can be applied to test the accuracy and predictive power of the model. The OECD has developed a set of rules that should be applied in the validation of QSPR and QSAR models.

References

Abraham, M.H. (1993). Scales of solute hydrogen-bonding - their construction and application to physicochemical and biochemical processes. Chemical Society Reviews 22, 73-83.

Armitage, J.M., Arnot, J.A., Wania, F., Mackay, D. (2013). Development and evaluation of a mechanistic bioconcentration model for ionogenic organic chemicals in fish. Environmental Toxicology and Chemistry 32, 115-128.

Bruggeman, W.A., Opperhuizen, A., Wijbenga, A., Hutzinger, O. (1984). Bioaccumulation of super-lipophilic chemicals in fish. Toxicological and Environmental Chemistry 7, 173-189.

Droge, S.T.J., Goss, K.U. (2013). Development and evaluation of a new sorption model for organic cations in soil: Contributions from organic matter and clay minerals. Environmental Science and Technology 47, 14233-14241.

Endo, S., Escher, B.I., Goss, K.U. (2011). Capacities of membrane lipids to accumulate neutral organic chemicals. Environmental Science and Technology 45, 5912-5921.

Endo, S., Goss, K.U. (2014). Applications of polyparameter linear free energy relationships in environmental chemistry. Environmental Science and Technology 48, 12477-12491.

Eriksson, L., Hermens, J.L.M., Johansson, E., Verhaar, H.J.M., Wold, S. (1995). Multivariate analysis of aquatic toxicity data with pls. Aquatic Sciences 57:217-241.

Geisler, A., Endo, S., Goss, K.U. (2012). Partitioning of organic chemicals to storage lipids: Elucidating the dependence on fatty acid composition and temperature. Environmental Science and Technology 46, 9519-9524.

Goss, K.-U., Schwarzenbach, R.P. (2001). Linear free energy relationships used to evaluate equilibrium partittioning of organic compounds. Environmental Science and Technology 35, 1-9.

Goss, K.U., Schwarzenbach, R.P. (2003). Rules of thumb for assessing equilibrium partitioning of organic compounds: Successes and pitfalls. Journal of Chemical Education 80, 450-455.

Hansch, C., Streich, M., Geiger, F., Muir, R.M., Maloney, P.P., Fujita, T. (1963). Correlation of biological activity of plant growth regulators and chloromycetin derivatives with hammett constants and partition coefficients. Journal of the American Chemical Society 85, 2817-&.

Jonker, M.T.O., van der Heijden, S.A. (2007). Bioconcentration factor hydrophobicity cutoff: An artificial phenomenon reconstructed. Environmental Science and Technology 41, 7363-7369.

Katritzky, A.R., Slavov, S., Radzvilovits, M., Stoyanova-Slavova, I., Karelson, M. (2009). Computational chemistry approaches for understanding how structure determines properties. Zeitschrift Fur Naturforschung Section B-a Journal of Chemical Sciences 64:773-777.

Könemann, H., Van Leeuwen, K. (1980). Toxicokinetics in fish: Accumulation and elimination of six chlorobenzenes by guppies. Chemosphere 9, 3-19.

Kraaij, R., Mayer, P., Busser, F.J.M., Bolscher, M.V., Seinen, W., Tolls, J. (2003). Measured pore-water concentrations make equilibrium partitioning work - a data analysis. Environmental Science and Technology 37, 268-274.

Sabljic, A., Güsten, H., Verhaar, H.J.M., Hermens, J.L.M. (1995). Qsar modelling of soil sorption. Improvements and systematics of log koc vs. Log kow correlations. Chemosphere 31, 4489-4514.

Tülp, H.C., Fenner, K., Schwarzenbach, R.P., Goss, K.U. (2009). pH-dependent sorption of acidic organic chemicals to soil organic matter. Environmental Science and Technology 43, 9189-9195.

Veith, G.D., Defoe, D.L., Bergstedt, B.V. (1979). Measuring and estimating the bioconcentration factor of chemicals in fish. Journal of the Fisheries Research Board of Canada 36, 1040-1048.

3.5. METAL SPECIATION

Introduction

Metals occur in different physical and chemical forms in the environment, for example as the element (very rare in the environment), as components of minerals, as free cations dissolved in water (e.g. Cd2+), or bound to inorganic or organic molecules in either the solid or dissolved phases (e.g. HgCH3+ or AgCl2+) (Allen 1993). The distribution of a metal over these different forms is referred to as metal speciation. Physical processes may also affect the mobility and bioavailability of metals, for example the electrostatic attraction of metal cations to negatively charged mineral surfaces. These processes are in general not referred to as metal speciation in the strict sense but they are discussed here.

Metal speciation Reactions

The speciation of metals is controlled by both the properties of the metals (see the section Metals and metalloids) and the properties of the environment in which they are present, such as the pH, redox potential and the presence and concentrations and properties of molecules that could form complexes with the metals. These complex forming molecules are often called ligands and these can vary from relatively simple anions in solution, such as sulphate or anions of soluble organic acids, to more complex macromolecules such as proteins and other biomolecules. The adsorption of metals by covalent bond formation to oxide and hydroxide surfaces of minerals, and oxygen- or nitrogen-containing functional groups of solid organic matter, is also referred to as complexation. Since these metal-binding functional groups are often either acidic or basic, the pH is an important environmental parameter controlling complexation reactions.

In natural systems the speciation of metals is of great complexity and determines their mobility in the environment and their bioavailability (i.e. how easily they are taken up by organisms). Metal speciation therefore plays a key role in determining the potential bioaccumulation and toxicity of metals and should therefore be considered when assessing their ecological risks. Metal bioavailability and transport are in particular strongly related to the distribution over solid and liquid phases of the environmental matrix.

The four main chemical reactions determining metal speciation, so the binding of metal ions to ligands and their presence in solid and liquid phases, are (Bourg, 1995):

- Adsorption and desorption processes
- Ion exchange and dissolution reactions
- Precipitation and co-precipitation
- Complexation to inorganic and organic ligands

The complexity of these reactions is illustrated in Figure 1.

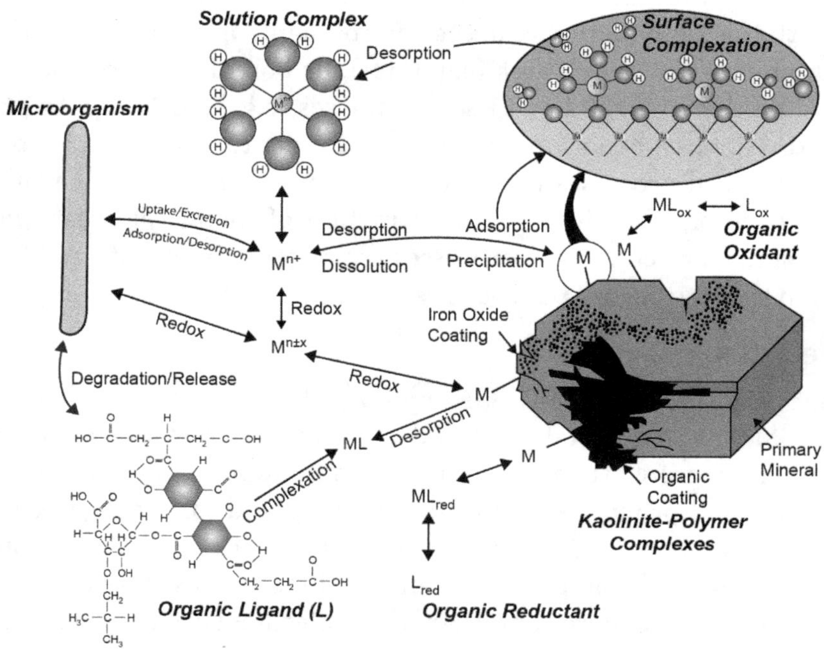

Figure 1. Metals (M) speciation in the environment is determined by a number of reactions, including complexation, precipitation and sorption. These reactions affect the partitioning of metals across solid and liquid phases, hence their mobility as well as their bioavailability. Drawn by Evelin Karsten-Meessen.

Adsorption, desorption and ion exchange processes take place with the reactive components present in soils, sediments and to lower extent in water. These include:

- Clay minerals
- Hydroxides (e.g. $FeOH_3$) and/or carbonates (e.g. $CaCO_3$)
- Organic matter

Metal ions react with these reactive components in different ways. In soils and sediments, cationic metals bind reversibly to clay minerals via cation-exchange processes (see section on Soil). Metal ions also form complexes with so-called functional groups (mainly carboxylic and phenolic groups) present in organic matter (see section on Soil). In aquatic systems similar binding processes occur, in which dissolved organic matter (or carbon) (DOM or DOC) plays a major role. The "dissolved" fraction of organic matter is operationally defined as the fraction passing a 0.45 µm filter and is often referred to fulvic and humic acids.

As mentioned above and in the section on Shttps://maken.wikiwijs.nl/147644/Environmental_Toxicology__an_open_online_textbook#!page-5415168oil, negatively charged surfaces of the reactive mineral and organic components present in soil, sediment or water attract positively-charged atoms or molecules (cations, e.g. Cd2+), and allow these cations to exchange with other positively charged ions. The competition between cations for binding sites is driven by the binding affinity of each metal species, as well as the concentration of each metal species. Cation-exchange capacity (CEC) is a property of the sorbent, and defined as the density of available negatively charged sites per mass of environmental matrix (soil, sediment). In fact, it is a measure of how many cations can be retained on solid surfaces. CEC usually is expressed in cmolc/kg soil (see section on Soil). Increasing the pH (i.e. decreasing the concentration of H+ ions) increases the variable charge of most sorbents (more types of protonated groups on sorbent surfaces release their H+), especially for organic matter, and therefore also increases the cation exchange capacity. Protons (H+) also compete with metal ions for the same binding sites. Conversely, at decreasing pH (increasing H+ concentrations), most sorbents lower their CEC.

Modelling Metal Speciation

Metal speciation can be modelled if we have sufficient knowledge of the most important reactions involved and the environmental conditions that control these reactions. This knowledge is expressed in the form of equilibria expressing the most important complexation and/or redox reactions. For example, in the general case of a complexation reaction between metal M and ligand L described by the equilibrium:

aM^{m+} *(aq)* + bL^{n-} *(aq)* Ã¢Â†Â" $M_aL_b{}^{q+}$ (where q = am-bn)

The relationship between the concentrations (or more accurately the activities) of the species is given by:

$$K_f = \frac{[M_a\, L_b^{q+}]}{[M^{m+}]^a + [L^{n-}]^b}$$

If we know the value of Kf, either from experimental measurements or by estimation, we can calculate the relative concentrations or activities of the free and complexed metal ions. The actual concentrations can be calculated if we either measure one of these concentrations and that of the ligand directly or measure the total concentration of the metal present (and of the ligand) and apply a mass balance model. The copper speciation in a salt water without any DOC, for instance, depends on pH as was described by Blust et al. (1991), see Figure 2. At pH 7.5, most Cu is bound to CO32-, but at pH 6 the Cu is mainly present as the free Cu2+ ion and in complexes with chloride (as CuCl+) and sulphate (as CuSO4).

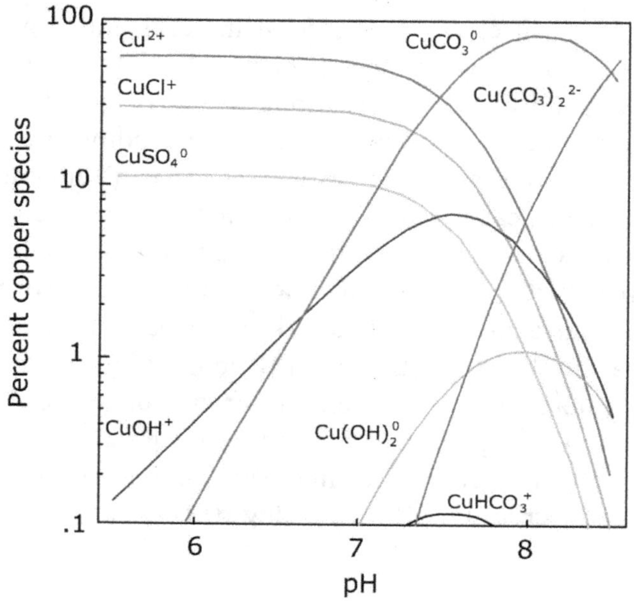

Figure 2. Copper speciation in salt water without DOC. Redrawn from Blust et al. (1991) by Wilma Ijzerman.

If redox reactions are involved in speciation, we can use the Nernst equation to describe the equilibrium between reduced and oxidised states of the metal:

$$E_h = E_h^0 - \frac{2.303RT}{nF} log \frac{(Red))}{(Ox)}$$

Where Eh is the redox potential, Eh0 the standard potential of the redox pair (relative to the hydrogen electrode), R the molar gas constant, T the temperature, n the number of transferred electrons, F the Faraday constant and {Red/Ox} the activity (or concentration) ratio of the reduced and oxidized species. Since many redox reactions involve the transfer of H+, the value of {Red/Ox} for these equilibria will depend on the pH. Note that the redox potential is often expressed as pe which is defined as the negative logarithm of the electron activity (pe = - log {e-}).

Using these comparatively simple equations for all the relevant reactions involved it is possible to construct models to describe metal speciation as a function of ligand concentrations, pH and redox potential. As an example, Table 1 presents the relevant equilibria for the speciation of iron in water.

Table 1. Equilibrium reactions relevant for Fe in water (adapted from Essington, 2003)

Boundary	Equilibrium reaction	
(1)	Fe3+ + e- D Fe2+	pEΘ = 13.05
(2)	Fe(OH)3(s) + 3H+ D Fe3+ + 3H2O	Ksp = 9.1 x 103 L2 mol-2
(3)	Fe(OH)2(s) + 2H+ D Fe2+ + 2H2O	K*sp = 8.0 x 1012 L mol-1
(4)	Fe(OH)3(s) + H+ + e- D Fe(OH)2(s) + H2O	
(5)	Fe(OH)3(s) + 3H+ + e- D Fe2+ + 3H2O	

Using these we can derive equations defining the conditions of pH and pe at which the activity or concentration ratio is one for each equilibrium. These are shown as the continuous boundary lines in Fig. 3. In this Pourbaix or pe-pH (or pE-pH) diagram, the fields separated by the boundary lines are labelled with the dominant species present under the conditions that define the fields. (NB. The dotted lines define the conditions of pe and pH under which water is stable.)

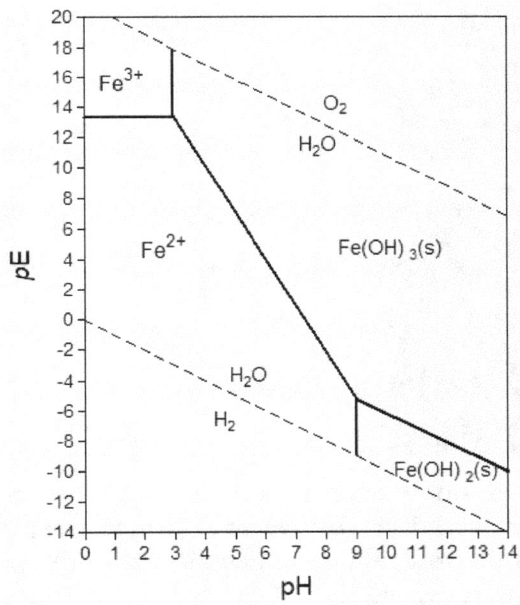

Figure 3. A pe-pH diagram for Fe in water showing the dominant species present under different conditions assuming a maximum soluble Fe(II) or (III) concentration of 10-5 mol L-1 (adapted from Essington, 2003).

Environmental Effects on Speciation

In the environment there is, however, in general no equilibrium. This means that the speciation and hence also fate of metals is highly dynamic. Large scale alterations occur when land use changes, e.g. when agricultural land is abandoned and becomes nature. Whereas agricultural soil often is 'limed' (addition of CaCO3) to maintain near-neutral pH and crop is removed by harvesting, in natural ecosystems all produced organic matter remains in the system. Therefore natural soils show an increase in soil organic matter content, while due to microbial decomposition processes soil pH tends to decrease. As a result, DOC concentration in the soil porewater will increase, while metal mobility also is increased by the decreasing soil pH (Cu2+ is more mobile than CuCO3). This may cause historical metal pollution to suddenly become available (the "chemical time bomb" effect). Large scale reconstruction of rivers or deep soil digging for land planning and development may also affect environmental conditions in such a way that metal speciation may change. An example of this is the change in arsenic speciation in groundwater due to the drilling of wells in countries like Bangladesh; the introduction of oxygen and organic matter into the deeper groundwater caused a change of arsenic speciation, enhancing its solubility in water and therefore increasing human exposure (see section on Metals and metalloids).

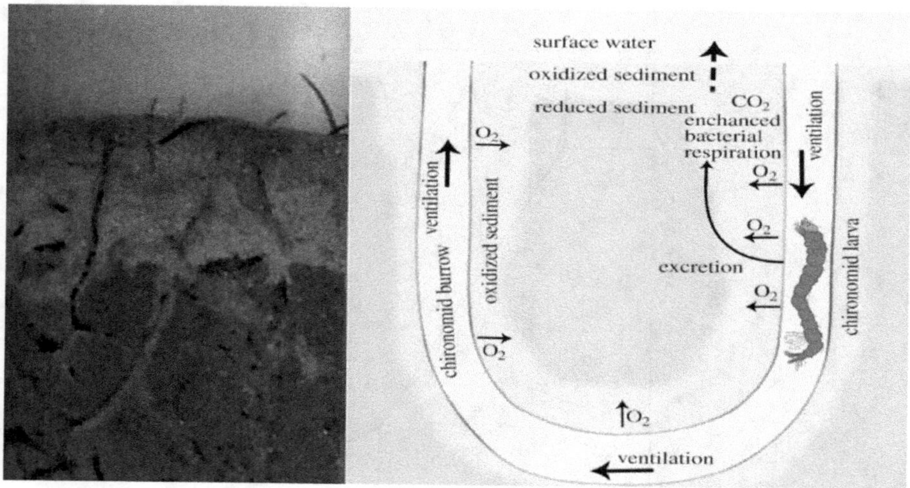

Figure 4. Illustration of small scale processes that alter metal speciation. (A) different bioturbation activities by various organisms, (B) the burrowing activities of a chironomid (midge) larvae can alter the environmental conditions and with that affect metal speciation in the sediment. Source: Martina Vijver.

Dynamic conditions do not only occur on large spatial and temporal scales, nature is also dynamic on smaller scales. Abiotic factors such as rain and flooding events, weather conditions, and redox status may alter metal speciation. In addition, biotic factors may affect metal speciation.

An example of the latter is the bioturbation by sediment-dwelling organisms that re-suspend particles into water, or earthworms that by their digging activities aerate the soil and excrete mucus that may stimulate microbial activity (see Figure

4A). These activities of soil and sediment organisms alter the environmental conditions and hence affect metal speciation (see Figure 4B). The production of acidic root exudates by plants may also have similar effects on metal speciation. Another process that alters metal speciation is the uptake of metals. Since the ionic metal form seems most prone to root uptake, or active intake over cell membranes, this process may affect metal partitioning over different species.

References

Allen, H.E. (1993). The significance of trace metal speciation for water, sediment and soil quality criteria and standards. Science of the Total Environment 134, 23-45.

Andrews, J.E., Brimblecombe, P., Jickells, T.D., Liss P.S., Reid, B. (2004) An Introduction to Environmental Chemistry, 2nd Edition, Blackwell, ISBN 0-632-05905-2 (chapter 6).

Bourg, A.C.M. (1995) Speciation of heavy metals in soils and groundwater and implications for their natural and provoked mobility. In: Salomons, W., Förstner, U., Mader, P. (Eds.). Heavy Metals. Springer, Berlin. p. 19-31.

Blust, R., Fontaine, A., Decleir, W. (1991) Effect of hydrogen ions and inorganic complexing on the uptake of copper by the brine shrimp Artemia franciscana. Marine Ecology Progress Series 76, 273-282.

Essington, M.E. (2003) Soil and Water Chemistry, CRC Press, ISBN 0-8493-1258-2 (chapters 5, 7 and 9).

Sposito, G. (2008) The Chemistry of Soils, 2nd Edition, Oxford University press, ISBN 978-0-19-531369-7 (chapter 4).

Sparks, D.L. (2003) Environmental Soil Chemistry, 2nd Edition, Academic Press, ISBN 0-12-656446-9 (chapters 5, 6 and 8).

3.6. AVAILABILITY AND BIOAVAILABILITY

3.6.1. Definitions

Introduction

Although many environmental chemists, toxicologists, and engineers claim to know what bioavailability means, the term eludes a consensus definition. Bioavailability may be defined as that fraction of chemical present in the environment that is or may become available for biological uptake by passage across cell membranes.

Bioavailability generally is approached from a process-oriented point of view within a toxicological framework, which is applicable to all types of chemicals (Figure 1).

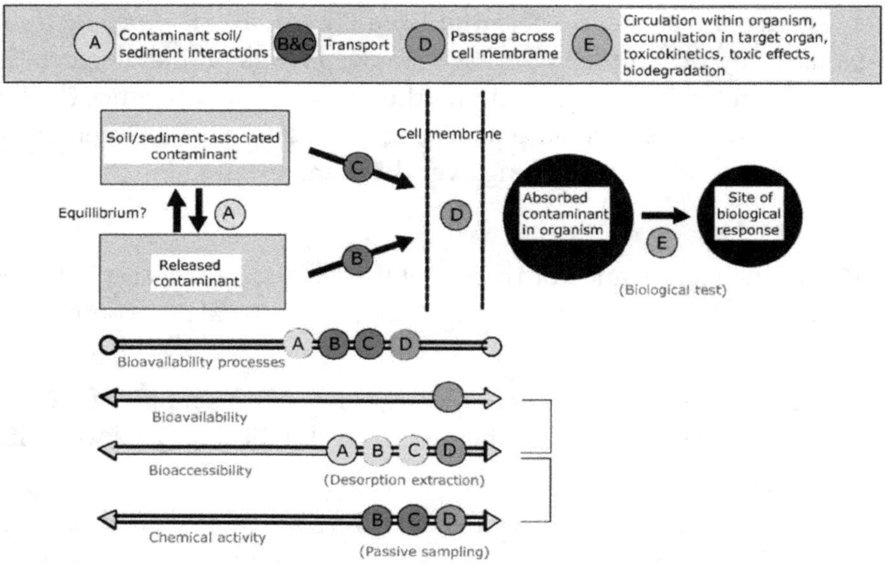

Figure 1. Bioavailability relates to a series of processes, ranging from processes external of organisms, towards internal tissues, and fully internal to the biological response site. Redrawn from Ortega-Calvo et al. (2015) by Wilma IJzerman.

The first process is chemical availability which can be defined as the fraction of the total concentration of chemicals present in an environmental compartment that contributes to the exposure of an organism. The total concentration in an environmental compartment is not necessarily involved in the exposure, as a smaller or larger fraction of the chemical may be bound to organic or inorganic components of the environment. Organic matter and clay particles, for instance, are important in binding chemicals (see section on Soil), while also the presence of cations and pH are important factors modifying the partitioning of chemicals between different environmental phases (see section on Metal speciation).

The second process is the actual or potential uptake, described as the toxicokinetics of a substance which reflects the development with time of its concentration on, and in, the organism (see section on Bioconcentration and kinetics modelling).

The third process describes the internal distribution of the substance leading to its interaction(s) at the cellular site of toxicity activation. This is sometimes referred to as toxico-availability and also includes the biochemical and physiological processes resulting from the effects of the chemical at the site of action.

Details on the bioavailability concept described above as well as how the physico-chemical interactions influencing each process are described in the sections on Metal speciation and Bioconcentration and kinetics modelling.

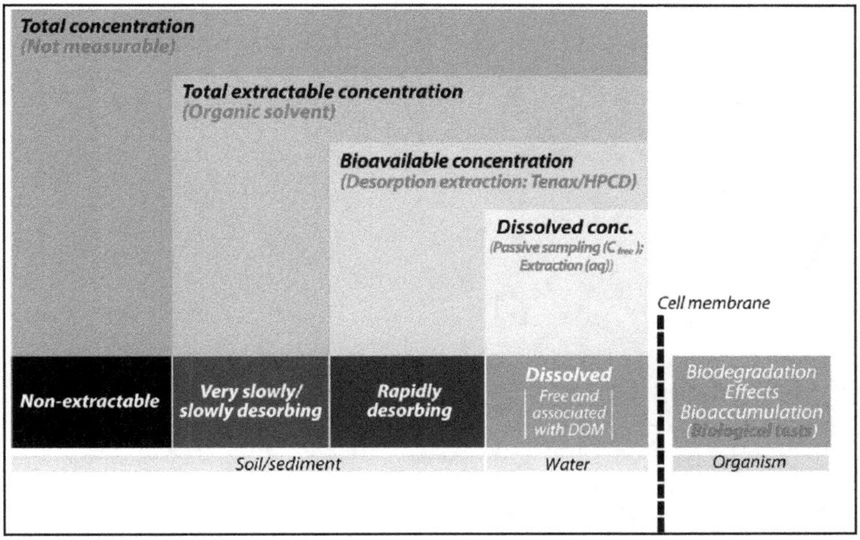

Figure 2. Bioavailability relates to a series of time frames, particularly in external processes (according to Ortega-Calvo et al. 2015).

Kinetics are involved in all three basic processes. The timeframe can vary from very brief (less than seconds) to very long in the order of hundreds of years. Figure 2 shows that some fractions of pollutants are present in soil or sediment, but may never contribute to the transport of chemicals that could reach the internal site during an organism's lifespan. The fractions with different desorption kinetics may relate to different experimental techniques to determine the relevant bioavailability metric.

Box 1: Illustration of how bioavailability influences our human fitness

Iron deficiency occurs when a body has not enough iron to supply its needs. Iron is present in all cells of the human body and has several vital functions. It is a key component of the hemoglobin protein, carrying oxygen to the tissues from the lungs. Iron also plays an important role in oxidation/reduction reactions, which are crucial for the functioning of the cytochrome P450 enzymes that are responsible for the biotransformation of endogenic as well as xenobiotic chemicals. Iron deficiency therefore can interfere with these vital functions, leading to a lack of energy (feeling tired) and eventually to malfunctioning of muscles and the brain.

In case of iron deficiency, the medical doctor will prescribe Fe-supplements and iron-rich food such as red meat and green leafy vegetables like spinach. Although this will lead to a higher intake of iron (after all exposure is higher), it does not necessarily lead to a higher uptake as here bioavailability becomes important. It is advised to avoid drinking milk or caffeinated drinks together with eating iron-rich products or supplements because both drinks will prevent the absorption of iron in the intestinal tract. Calcium ions abundant in milk will compete with iron ions for the same uptake sites, so excess calcium will reduce iron uptake. Carbonates and caffeine molecules, but also phytate (inositol polyphosphate) present in vegetables, will strongly bind the iron, also reducing its availability for uptake.

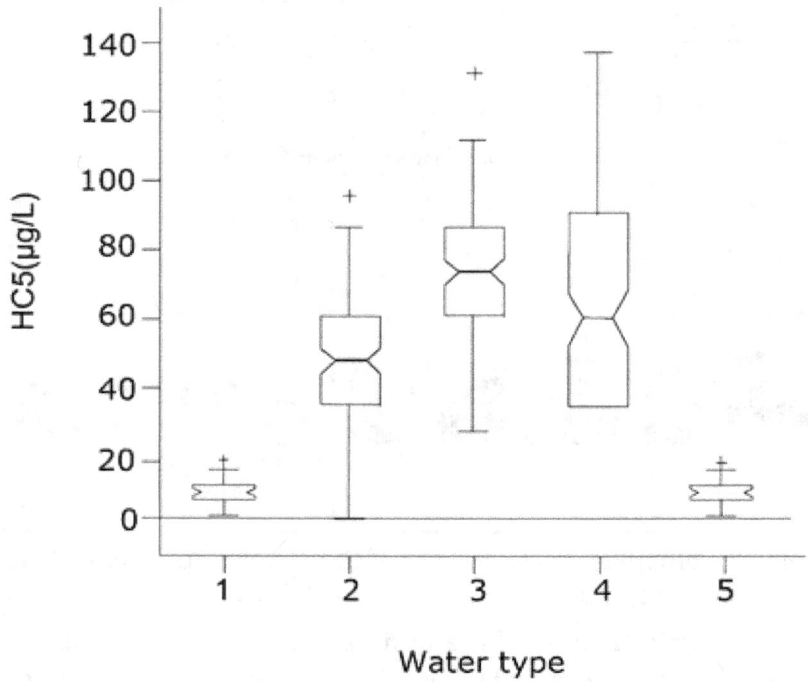

Figure 3. Bioavailability correction to estimate the HC_5 copper concentration in relation to properties like dissolved organic carbon content and pH (Table 1) in order to estimate its risk in different water types (according to Vijver et al. 2008), in comparison to the 1.5 µg/L total dissolved Cu in surface waters as the current generic Dutch standard (horizontal line). Redrawn from Vijver et al. (2008) by Wilma Ijzerman

Bioavailability used in Risk Assessment

For regulatory purposes, it is necessary to use a straightforward approach to assess and prioritize contaminated sites based on their risk to human and environmental health. The bioavailability concept offers a scientific underpinned concept to be used in risk assessment. Examples for inorganic contaminants are the derived 2nd tier models such as the Biotic Ligand Models, while for organic chemicals the Equilibrium Partitioning (EqP) concept (see Box 2 in the section on Sorption) is applied.

A quantitative example is given for copper in different water types in Figure 3 and Table 1, in which water chemistry is explicitly accounted for to enable estimating the available copper concentration. The current Dutch generic quality target for surface waters is 1.5 µg/L total dissolved copper. The bioavailability-corrected risk limits (HC_5) for different water types, in most cases, exceeded this generic quality target.

Table 1. Bioavailability adjusted Copper 5% Hazardous Concentration (HC5, potentially affecting <5% of relevant species) for different water types.

Water type description	No.	DOC (mg/L)	pH	Average HC5 (µg/L)
Large rivers	1	3.1 ± 0.9	7.7 ± 0.2	9.6 ± 2.9
Canals, lakes	2	8.4 ± 4.4	8.1 ± 0.4	35.0 ± 17.9
Streams, brooks	3	18.2 ± 4.3	7.4 ± 0.1	73.6 ± 18.9
Ditches	4	27.5 ± 12.2	6.9 ± 0.8	64.1 ± 34.5
Sandy springs	5	2.2 ± 1.0	6.7 ± 0.1	7.2 ± 3.1

When the calculated HC5 value is lower, this means that the bioavailability of copper is higher and hence at the same total copper concentration in water the risk is higher. The bioavailability-corrected HC5s for Cu differ significantly among water types. The lowest HC5 values were found for sandy springs (water type V) and large rivers (water type I), which appear to be sensitive water bodies. These differences can be explained from partitioning processes (chemical availability) and competition processes (the toxicokinetics step) on which the BLMs are based. Streams and brooks (water type III) have rather high total copper concentrations without any adverse effects, which can be attributed to the protective effect of relatively high dissolved organic carbon (DOC) concentrations and the neutral to basic pH causing a high binding of Cu to the DOC.

For risk managers, this water type specific risk approach can help to identify the priority in cleanup activities among sites having elevated copper concentrations. It remains possible that, for extreme environmental situations (e.g., extreme droughts and low water discharges or extreme rain fall and high runoff), combinations of the water chemistry parameters may result in calculated HC5 values that are even lower than the calculated average values. For the latter (important) reason, the generic quality target is more strict.

References

Hamelink, J., Landrum, P.F., Bergman, H., Benson, W.H. (1994) Bioavailability: physical, chemical, and biological interactions, CRC Press.

Ortega-Calvo, J.J., Harmsen, J., Parsons, J.R., Semple, K.T., Aitken, M.D., Ajao, C., Eadsforth, C., Galay-Burgos, M., Naidu, R., Oliver, R., Peijnenburg, W.J.G.M., Römbke, J., Streck, G., Versonnen, B. (2015) From bioavailability science to regulation of organic chemicals. Environmental Science and Technology 49, 10255–10264.

Vijver, M.G., de Koning, A., Peijnenburg, W.J.G.M. (2008) Uncertainty of water type-specific hazardous copper concentrations derived with biotic ligand models. Environmental Toxicology and Chemistry 27, 2311-2319.

3.6.2. Assessing Available Concentrations of Organic Chemicals

Introduction: Bioavailability through the Water Phase

In many exposure scenarios involving organic chemicals, ranging from a bacterial cell to a fish, or from a sediment bed to a soil profile, the organisms experience the pollution through the water phase. Even when this is not the case, for example when uptake is from sediment consumed as food, the aqueous concentration may be a good indicator of the bioavailable concentration, since ultimately a chemical equilibrium will be established between the solid phase, the aqueous phase (possibly in the intestine), and the organism. Thus, taking an aqueous sample from a given environment, and determining the concentration of a certain chemical with the appropriate analytical equipment seems a straightforward approach to assess bioavailability. However, especially for hydrophobic chemicals, which tend to remain sorbed to solid surfaces (see sections on Relevant chemical properties and Sorption of organic chemicals), the determination of the chemicals present in the aqueous phase, as a way to assess bioavailability, has represented a significant challenge to environmental organic chemistry.

The phase exchange among different compartments often leads to equilibrium aqueous concentrations that are very low, because most of the chemicals remain associated to the solids, and after sustained exposure, to the biota. These freely dissolved concentrations (Cfree) are very useful to determine bioavailability, as they represent the "tip of the iceberg" under equilibrium exposure, and are what organisms "see" (Figure 1, left). Similarly to the balance between gravity and buoyancy forces leading to iceberg flotation up to a certain level, Cfree is determined by the equilibrium between sorption and desorption, and connected to the concentration of the sorbed chemical (Csorbed) through a partitioning coefficient.

Biological uptake may also result in the fast removal of the chemical from the aqueous phase, and thus in further desorption from the solids, so equilibrium is never achieved, and actual aqueous concentrations are much lower than the equilibrium Cfree (or even close to zero). In these situations, bioavailability is driven by the desorption kinetics of the chemical. Usually, desorption occurs as a biphasic process, where a fast desorption phase, occurring during a few hours or days, is followed by a much slower phase, taking months or even years. Therefore, for scenarios involving rapid exposures, or for studies on coupled desorption/biodegradation, the fast-desorbing fraction of the chemicals (Ffast) can be used to determine bioavailability. This fraction is often referred to as the bioaccessible fraction. Following the iceberg analogy (Figure 1, right), Ffast would constitute the upper iceberg fraction rapidly melting by sun irradiation, with a very minimal "visible" surface (representing the desorbed chemical in the aqueous solution, which is quickly removed by biological uptake). The slowly desorbing –or melting- fraction, Fslow, would remain in the sorbed state, within a given time span, having little interactions with the biota.

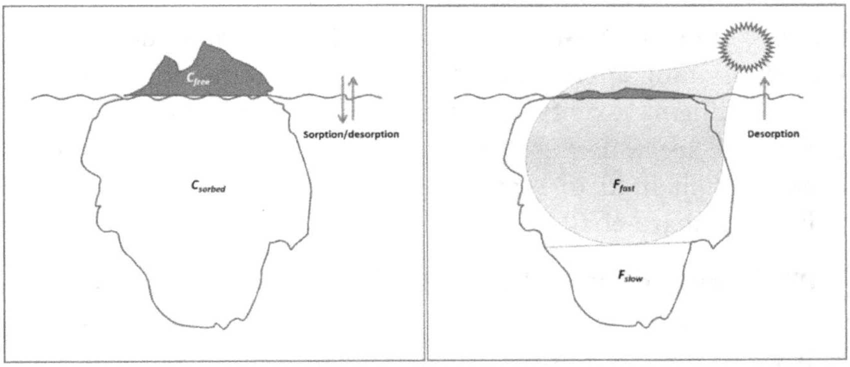

Figure 1. The magnitude of Cfree, determined by the sorption/desorption equilibrium (similarly to a floating iceberg, left), can correspond to a minimal fraction of the total pollutant mass, but may constitute the main driver for bioavailability (and risk) in equilibrium exposure scenarios. In non-equilibrium conditions (right, in analogy, a melting iceberg exposed to irradiation to sun), the fraction of sorbed chemical that can be rapidly mobilized, Ffast, can be taken as an estimate of bioavailability.

Determining Bioavailability with Passive Sampling Methods

C_{free} can be determined with a passive sampler, in the form of polymer-coated fibers or sheets (membranes) made of a variety of polymers, which establish an additional sorption equilibrium with the aqueous phase in contact with the soil or sediment (Jonker et al., 2018). Depending on the analytes of interest, different polymers, such as polydimethylsiloxane (PDMS) or polyethylene (PE), are used in passive samplers. The passive sampler, enriched in the analyte (similarly to the floating iceberg in Figure 1, left, where Csorbed in this case is the concentration in the passive sampler), can be used in this way to determine indirectly the pollutant concentration present in the aqueous phase, even at very low concentrations, though the appropriate distribution ratio between sampler and water. In bioavailability estimations, passive sampling is designed for equilibrium and non-depletive conditions. This means that the amount of chemical sampled does not alter the solid-water equilibrium, i.e., it is essential that Cfree is not affected significantly by the sampler. Equilibrium achievement is critical, and it may take days or weeks.

C_{free} can be calculated from the concentration of the pollutant in the passive sample polymer at equilibrium (Cp), and the polymer-to-water partitioning coefficient (Kpw):

$$C_{free} = \frac{C_p}{K_{pw}}$$

C_{free} values can be the basis of predictions for bioaccumulation that use the equilibrium partitioning approach, either directly or through a bioconcentration factor, and for sediment toxicity in conjunction with actual toxicity tests. Passive sampling methods are well suited for contaminated sediments, and they have already been implemented in regulatory environmental assessments based on bioavailability (Burkhard et al., 2017).

Determining Bioavailability with Desorption Extraction Methods

The determination of F_{fast} can be achieved with the use of methods that trap the desorbed chemical once it appears in the aqueous phase. Far from equilibrium conditions, desorption is driven to its maximum rate by placing a material in the aqueous phase that acts as an infinite sink (comparable to the sun irradiation of a melting iceberg in Figure 1, right).

The most accepted materials for these desorption extraction methods are Tenax, a sorptive resin, and cyclodextrin, a solubilizing agent (ISO, 2018). These methods allow a permanent aqueous chemical concentration of almost zero, and therefore, sorption of the chemical back to the soil or sediment can be neglected. Several extraction steps can be used, covering a variable time span, which depends on the environmental sample.

The following first-order, two-compartment kinetic model can be used to analyze desorption extraction data:

$$\frac{St}{S0} = F_{fast}\, e^{-k_{fast}\, t} + F_{slow}\, e^{-k_{slow}\, t}$$

In this equation, St and So (mg) are the soil-sorbed amounts of the chemical at time t (h) and at the start of the experiment, respectively. Ffast and Fslow are the fast- and slow-desorbing fractions, and kfast and kslow (h-1) are the rate constants of fast and slow desorption, respectively. To calculate the values of the different constants and fractions (Ffast, Fslow, kfast, and kslow) exponential curve fitting can be used. The ln form of the equation can be used to simplify curve fitting.

Once the desorption kinetics are known, the method can be simplified for a series of samples, by using single time point-extractions. A time period of 20 h has been suggested as a sufficient time period to approximate Ffast. It is highly convenient for operational reasons (ISO, 2018), but indicative at best, since the time needed to extract Ffast tends to vary between chemicals and soils/sediments.

References

Burkhard, L.P., Mount, D.,R., Burgess, R.,M. (2017). Developing Sediment Remediation Goals at Superfund Sites Based on Pore Water for the Protection of Benthic Organisms from Direct Toxicity to Nonionic Organic Contaminants EPA/600/R 15/289; U.S. Environmental Protection Agency Office of Research and Development: Washington, DC.

ISO (2018). Technical Committee ISO/TC 190 Soil quality — Environmental availability of non-polar organic compounds — Determination of the potentially bioavailable fraction and the non-bioavailable fraction using a strong adsorbent or complexing agent; International Organization for Standardization: Geneva, Switzerland.

Jonker, M.T.O., van der Heijden, S.A., Adelman, D., Apell, J.N., Burgess, R.M., Choi, Y., Fernandez, L.A., Flavetta, G.M., Ghosh, U., Gschwend, P.M., Hale, S.E., Jalalizadeh, M., Khairy, M., Lampi, M.A., Lao, W., Lohmann, R., Lydy, M.J., Maruya, K.A., Nutile, S.,A., Oen, A.M.P., Rakowska, M.I., Reible, D., Rusina, T.P., Smedes, F., Wu, Y. (2018) Advancing the use of passive sampling in risk assessment and management of sediments contaminated with hydrophobic organic chemicals: results of an international ex situ passive sampling interlaboratory comparison. Environmental Science & Technology 52 (6), 3574-3582.

3.6.3. Assessing Available Metal Concentrations

Introduction

Total concentrations are not very informative about the availability of metals in soils or sediments. Fate and behavior of metals - in general terms mobility - as well as biological uptake and toxicity is highly determined by their speciation. Speciation describes the partitioning of a metal among the various forms in which it may exist (see section on Metal speciation). For assessing the risk of metals to man and the environment, speciation therefore is highly relevant as it may determine their availability for uptake and effects in organisms.

Several tools have been developed to determine available metal concentrations or their speciation in soils and sediments. As indicated in the section on Availability and bioavailability, such chemical methods are just indicative, and to a large extent ignore dynamics of availability. Moreover, availability is also influenced by biological processes, with abiotic-biotic interactions influencing the bioavailability process being species- and often even life-stage specific. Nevertheless, chemical extractions may provide useful information to predict or estimate the potential risks of metals and therefore are preferred over the determination of total metal concentrations.

The available methods include:

- Porewater extraction
- Extractions with water
- Extractions with diluted salts
- Extractions with chelating agents
- Extractions with diluted acids
- Sequential extractions using a series of different extraction solutions
- Passive sampling methods

Porewater extraction probably best approaches the readily available fraction of metals in soil, which drives mobility and is the fraction of metals experienced directly by many organisms exposed. In general, pore water is extracted from soil or sediment by centrifugation, and filtration over a 0.45 μm (or 0.22 μm) filter to remove larger particles and perhaps some of the dissolved organic matter. Filtration, however, will not remove all complexes, making it impossible to determine solely the dissolved metal fraction in the pore water. Nevertheless, porewater metal concentrations have been shown to have significant correlations with metal uptake (e.g. for copper uptake by barley and tomato by Zhao et al., 2006) and to be useful for predicting toxic threshold concentrations of metals, with correction for pH (e.g. for nickel toxicity to tomato and barley by Rooney et al., 2007).

Extraction with water simulates the immediately available fraction, so the fraction present in the soil solution or pore water. By extracting soil with water, the pore water however, is diluted, which on one hand may facilitate metal analysis by creating larger volumes of solution, but on the other hand may lead to differences between measured and actual metal concentrations in the pore water as it may impact chemical equilibria.

Extraction with diluted salts aims to determine the fraction of metal that is easily available or may become available as it is in the exchangeable form. This refers to cationic metals that may be bound to the negatively charged soil particles (see section on Soil). Buffered salt solutions, for instance 1 M NH4-acetate at pH 4.8 (with acetic acid) or at pH 7, may under- or overestimate available metal concentrations because of their interference with soil pH. Unbuffered salt solutions therefore are more widely used and may for instance include 0.001 or 0.01 M CaCl2, 0.1 M NaNO3 or 1 M NH_4NO_3 (Gupta and Aten, 1993; Novozamsky et al., 1993). Gupta and Aten (1993) showed good correlations between the uptake of some metals in plants and 0.1 M NaNO3 extractable concentrations in soil, while Novozamsky et al. (1993) found similar well-fitting correlations using 0.01 M $CaCl_2$. The latter method also seemed well capable of predicting metal uptake in soil invertebrates, and therefore has been more widely accepted for predicting metal availability in soil ecotoxicology. Figure 1 (Zhang *et al.*, 2019) provides an example with the correlation between Pb toxicity to enchytraeid worms in different soils and 0.01 M CaCl2 extractable concentrations.

Extractions with water (including porewater) and dilute salts are most accurately described as measures of the chemical solubility of the metal in the soil. The values obtained can be useful indicators of the relative metal reactivity across soils, but tend to be less useful for bioavailability assessment, unless the soils under consideration have a narrow range of soil properties. This is because the solutions obtained from such soils themselves have varying chemical properties (e.g. pH, DOC concentration) which are likely to affect the availability of the measured metal to organisms.

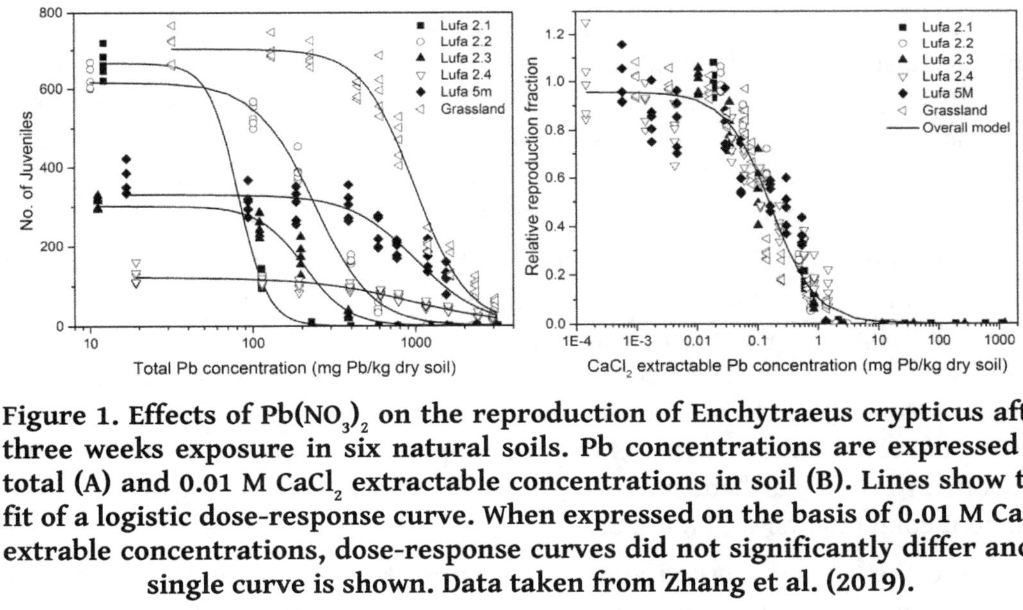

Figure 1. Effects of $Pb(NO_3)_2$ on the reproduction of Enchytraeus crypticus after three weeks exposure in six natural soils. Pb concentrations are expressed as total (A) and 0.01 M $CaCl_2$ extractable concentrations in soil (B). Lines show the fit of a logistic dose-response curve. When expressed on the basis of 0.01 M $CaCl_2$ extrable concentrations, dose-response curves did not significantly differ and a single curve is shown. Data taken from Zhang et al. (2019).

Extraction with chelating agents, such as EDTA (0.01-0.05 M) or DTPA (0.005 M) (as their sodium or ammonium salts), aims at assessing the availability of metals for plants. Many plants have the ability to actively affect metal speciation in the soil by producing root exudates. These extractants may form very stable water-soluble complexes with many different polyvalent cationic metals. It should be noted that the large variation in plant species and corresponding physiologies as well as their interactions with symbiotic microorganisms (e.g. mycorrhizal fungi) make that there is no single extraction method is capable of properly predicting metal availability to all plant species.

Extraction with diluted acids has been advocated for predicting the potentially available fraction of metals in soils, so the fraction that may become available in the long run. It is a quite rigorous extraction method that can be executed in a robust way. Metal concentrations determined by extracting soils with 0.43 M HNO_3 showed very good correlation with oral bioaccessible concentrations (Rodrigues et al., 2013), probably because it to some degree simulates metal release under acidic stomach conditions.

Both extraction methods with chelating agents and diluted acid may also dissolve solids, such as carbonates and Fe- and Al-oxides. This raises concerns as to the interpretation of results of these extraction systems, and especially to their generalization to different soil-plant systems (Novozamsky *et al.*, 1993). The extractions with chelating agents and dilute acids are considered methods to estimate the 'geochemically active' metal in soil - the pool of adsorbed metal that can participate in solid-solution adsorption/desorption and exchange equilibria on timescales of hours to days. This pool, along with the basic soil properties such as pH etc., also controls the readily available concentrations obtained with water/weak salt/porewater extraction. From the bioavailability point of view, these extractions tend to be most useful as inputs to bioavailability/toxicity models such as that of Lofts et al. (2014), which take further account of the effects of metal speciation and soil chemistry on metal bioavailability to environmental organisms.

Sequential extraction brings together different extraction methods, and aims to determining either how strongly metals are retained or to which components of the solid phase they are bound in soils or sediments. This allows to determine how metals are bound to different fractions within the same soil or sediment, and allows interpretation to the bioavailability dynamics. By far the most widely used method of sequential extraction is the one proposed by Tessier et al. (1979). Five fractions are distinguished, indicating how metals are interacting with soil or sediment components: see Figure 2.

Where the Tessier method aims at assessing the environmental availability of metals in soils and sediments, similar sequential extraction methods have also been developed for assessing the potential availability of metals for humans (bioaccessibility) following gut passage of soil particles (see e.g. Basta and Gradwohl, 2000).

Passive sampling may also be applied to assess available metal concentrations. The best known method is that of Diffusive Gradients in Thin films (DGT), developed by Zhang et al. (1998). In this method, a resin (Chelex) with high affinity for metals is placed in a device and covered with a diffusive gel and a 0.45 µm cellulose nitrate membrane (Figure 3). The membrane is brought into contact with the soil. Metals dissolved in the soil solution will diffuse through a membrane and diffusive gel and bind to the resin. Based on the thickness of the membrane and gel and the contact time with the soil, the metal concentration in the pore water can be calculated from the amount of metal accumulated in the resin. The method may be indicative of available metal concentrations in soils and sediments, but can only work effectively when soil is sufficiently moist to guarantee optimal diffusion of metals to the resin. For the same reasons, the method probably is better suited for assessing the availability of metals to plants than to invertebrates, especially for animals that are not in continuous contact with the soil solution.

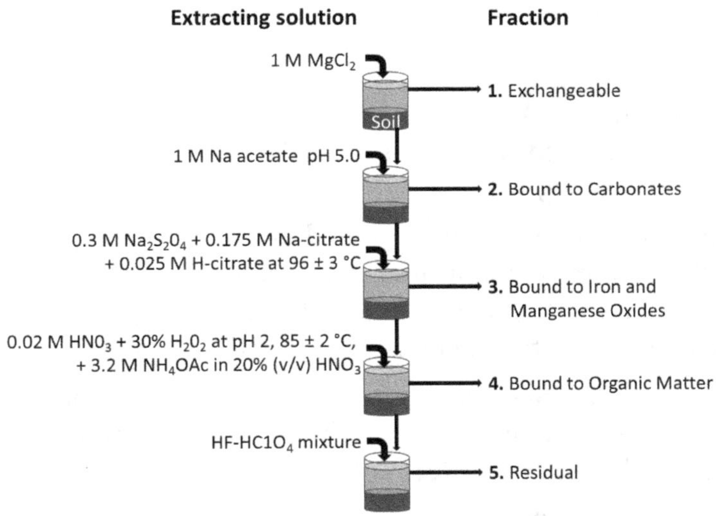

Figure 2. Schematic presentation of the sequential extraction of soil or sediment samples following the method of Tessier et al. (1979). The fractions obtained give an indication of the sites where metals are bound in the soil or sediment, and represent also an increasing binding strength, going from exchangeable to residual. Source: Kees van Gestel.

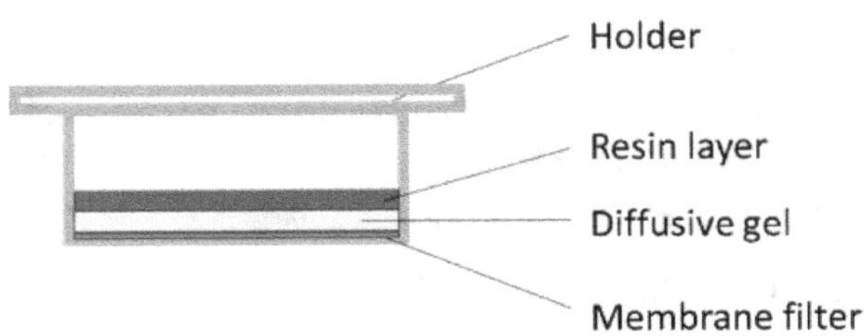

Figure 3. Device used in the Diffusive Gradients in Thin film (DGT) method for determining available metal concentrations in soil and sediment (adapted from Zhang et al., 1998). The device is placed on the soil or sediment in such a way that the membrane filter makes contact with the porewater. Metals may diffuse from the porewater to the resin layer. See text for further explanation.

Several of the above described methods have been adopted by the International Standardization Organization (ISO) in (draft) standardized test guidelines for assessing available metal fractions in soils, sediments and waste materials, e.g. to assess the potential for leaching to groundwater or their potential bioaccessibility. This includes e.g. ISO/TS 21268-1 (2007) "Soil quality - Leaching procedures for subsequent chemical and ecotoxicological testing of soil and soil materials - Part 1: Batch test using a liquid to solid ratio of 2 l/kg dry matter", ISO 19730 (2008) "Soil quality -Extraction of trace elements from soil using ammonium nitrate solution" and ISO 17586 (2016) "Soil quality -- Extraction of trace elements using dilute nitric acid".

References

Basta, N., Gradwohl, R. (2000). Estimation of Cd, Pb, and Zn bioavailability in smelter-contaminated soils by a sequential extraction procedure. Journal of Soil Contamination 9, 149-164.

Gupta, S.K., Aten, C. (1993). Comparison and evaluation of extraction media and their suitability in a simple model to predict the biological relevance of heavy metal concentrations in contaminated soils. International Journal of Environmental Analytical Chemistry 51, 25-46.

Lofts, S., Spurgeon, D.J., Svendsen, C., Tipping, E. (2004). Deriving soil critical limits for Cu, Zn, Cd, and Pb: A method based on free ion concentrations. Environmental Science and Technology 38, 3623-3631.

Novozamsky, I., Lexmond, Th.M., Houba, V.J.G. (1993). A single extraction procedure of soil for evaluation of uptake of some heavy metals by plants. International Journal of Environmental Analytical Chemistry 51, 47-58.

Rodrigues, S.M., Cruz, N., Coelho, C., Henriques, B., Carvalho, L., Duarte, A.C., Pereira, E., Römkens, P.F. (2013). Risk assessment for Cd, Cu, Pb and Zn in urban soils: chemical availability as the central concept. Environmental Pollution 183, 234-242.

Rooney, C.P., Zhao, F.-J., McGrath, S.P. (2007). Phytotoxicity of nickel in a range of European soils: Influence of soil properties, Ni solubility and speciation. Environmental Pollution 145, 596-605.

Tessier, A., Campbell, P.G.C., Bisson, M. (1979). Sequential extraction procedure for the speciation of particulate trace metals. Analytical Chemistry 51, 844-851.

Zhang, H., Davison, W., Knight, B., McGrath, S. (1998). In situ measurements of solution concentrations and fluxes of trace metals in soils using DGT. Environmental Science and Technology 32, 704-710.

Zhang, L., Verweij, R.A., Van Gestel, C.A.M. (2019). Effect of soil properties on Pb bioavailability and toxicity to the soil invertebrate Enchytraeus crypticus. Chemosphere 217, 9-17.

Zhao, F.J., Rooney, C.P., Zhang, H., McGrath, S.P. (2006). Comparison of soil solution speciation and diffusive gradients in thin-films measurement as an indicator of copper bioavailability to plants. Environmental Toxicology and Chemistry 25, 733-742.

3.7. DEGRADATION

3.7.1. Chemical and Photochemical Degradation Processes

Introduction

Transformation of organic chemicals in the environment can occur by a variety of reactions. These may be purely chemical reactions, such as hydrolyses or redox reactions, photochemical reactions with the direct or indirect involvement of light, or biochemical reactions.

Such transformations can change the biological activity (toxicity) of a molecule; it can change the physico-chemical properties and thus change its environmental partitioning processes; it can change its bioavailability, for example facilitating biodegradation; or it may contribute to the complete removal (mineralization) of the chemical from the environment.

In many cases, chemicals may be removed by combinations of these different processes and it is sometimes difficult to unequivocally identify the contributions of the different mechanisms. Indeed, combinations of different mechanisms are sometimes important, for example in cases where microbial activity is responsible for creating conditions that favour chemical reactions.

Here we will focus on two types of reactions: Abiotic (dark) reactions and photochemical reactions. Biodegradation reactions are covered elsewhere (see section on Biodegradation).

Chemical Degradation

Hydrolytic reactions are important chemical reactions removing organic contaminants and are particularly important for chemicals containing acid derivatives as functional groups. Common examples of such chemicals are pesticides of the organophosphate and carbamate classes such as parathion, diazinon, aldicarb and carbaryl.

Organophosphate chemicals are also used as flame retardants and are widely distributed in the environment. Some examples of hydrolysis reactions are shown in Figure 1.

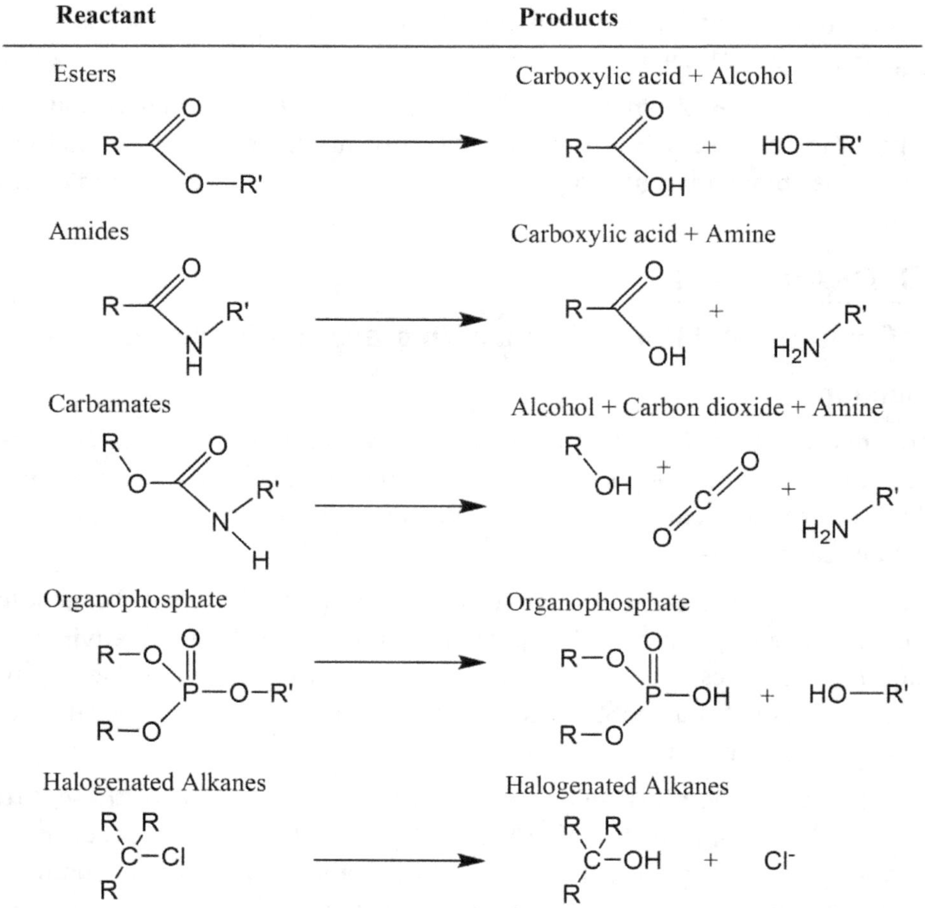

Figure 1 Examples of hydrolyses of esters and carbamates (redrawn after Van Leeuwen and Vermeire, 2007).

As the name suggests, hydrolysis reactions involve using water (hydro-) to break (-lysis) a bond. Hydrolyses are reactions with water to produce an acid and either an alcohol or amine as products. Hydrolyses can be catalysed by either OH- or H+ ions and their rates are therefore pH dependent. Some examples of pH-dependent ester hydrolysis reactions are shown in Figure 2.

Halogenated organic molecules may also be hydrolysed to form alcohols (releasing the halogen as a halide ion). The rates of these reactions vary strongly depending on the structure of the organohalogen molecule and the halogen substituent (with Br and I being substituted more rapidly than Cl, and much more rapidly than F) and in general the rates of these reactions are too slow to be of more than minor importance except for tertiary organohalogens and secondary organohalogens with Br and I (Schwarzenbach et al. 2017).

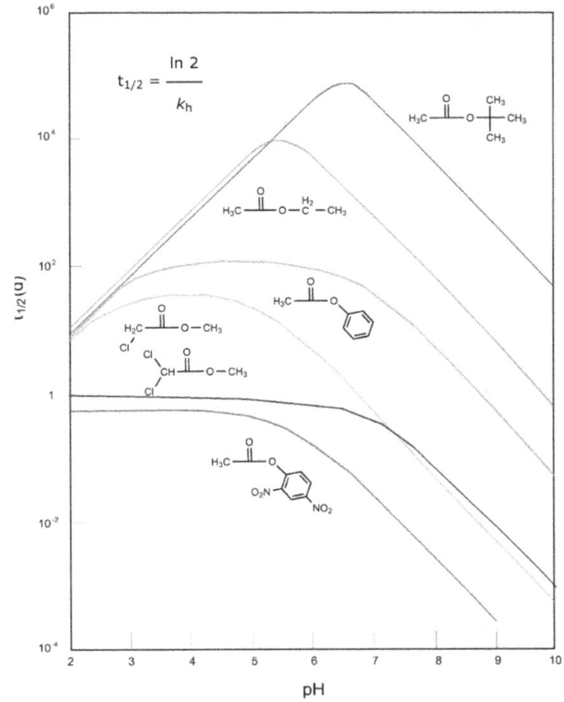

Figure 2. Examples of pH dependent ester hydrolysis reactions (Schwarzenbach et al. 2017). Note that the y-axis is half-life (on a logarithmic scale), meaning high values correspond to slow reactions. Redrawn by Wilma Ijzerman.

In some cases, other substitution reactions not involving water as reactant may be important. Some examples include Cl– in seawater converting CH_3I to CH_3Cl and reaction of thiols with alkyl bromines in anaerobic groundwater and sediment porewater under sulfate-reducing conditions (Schwarzenbach et al. 2017)

Redox (reduction and oxidation) reactions are another important reaction class involved in the degradation of organic chemicals. In the presence of oxygen, the oxidation of organic chemicals is thermodynamically favourable but occurs at insignificant rates unless oxygen is activated in the form of oxygen radicals or peroxides (following light absorption for example, see below) or if the reaction is catalysed by transition metals or transition metal-containing enzymes (see the sections on Biodegradation and Xenobiotic metabolism and defence).

Reduction reactions are important redox reactions for environmental contaminants in anaerobic environments such as sediment and groundwater aquifers. Under these conditions, organic chemicals containing reducible functional groups such as carboxylic acids and nitro groups undergo reduction reactions (Table 1).

Table 1: Examples of chemical redox reactions that may occur in the environment (adapted from Schwarzenbach et al. 2017)

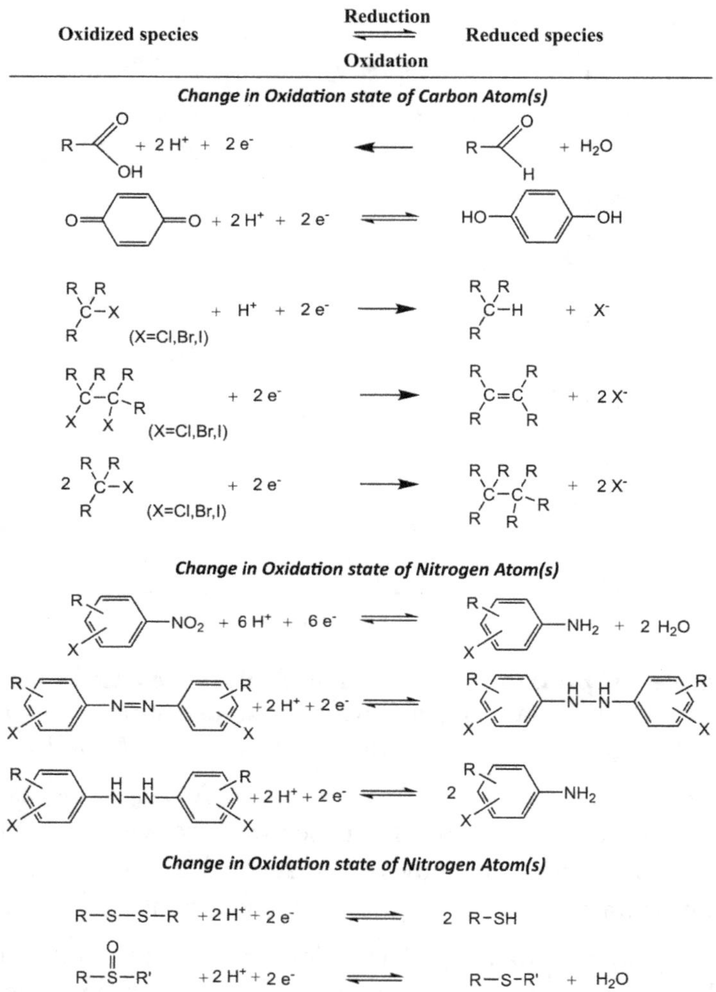

Organohalogens may also undergo reductions reactions with hydrogen where halogen substituents are replaced by hydrogen. These reactions are referred to as reductive dehalogenations and electron donors in these reaction can be inorganic oxidation reactions (such as the oxidation of Fe(II) minerals) or biochemical oxidation of organic chemicals. In fact, biological processes are also involved indirectly as the environmental redox conditions which determine which redox reactions can take place are in turn determined by microbial activity. Natural organic matter is often involved in environmental redox reactions as a catalyst enhancing electron transfer (Schwarzenbach et al. 2017). As an example, Figure 3 shows reductive dehalogenation reactions of hexachlorobenzene.

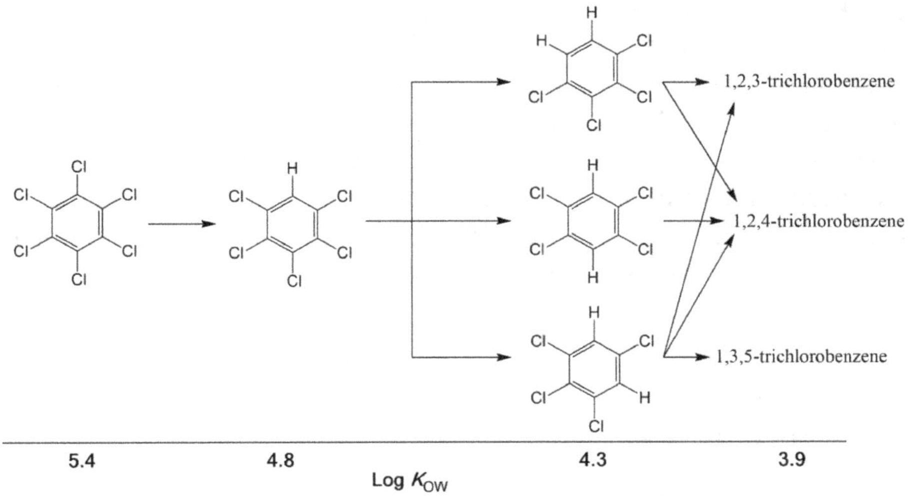

Figure 3. Reductive dehalogenation of hexachlorobenzene to less hydrophobic dechlorinated products (redrawn after Van Leeuwen and Vermeire, 2007).

Photodegradation

Sunlight is an important source of energy to initiate chemical reactions and photochemical reactions are particularly important in the atmosphere. Aromatic compounds and other chemicals containing unsaturated bonds that are able to absorb light in the frequency range available in sunlight become exited (energized) and this can lead to chemical reactions. These reactions lead to cleavage of bonds between carbon atoms and other atoms such as halogens to produce radical species. These radicals are highly reactive and react further to remove hydrogen or OH radicals from water to produce C-H or C-OH bonds or may react with themselves to produce larger molecules. Well known examples of atmospheric photochemical stratospheric reactions of CFCs that have had a negative impact on the so-called ozone layer and photochemical oxidations of hydrocarbons that are involved in the generation of smog.

In the aquatic environment, light penetration is sufficient to lead to photochemical reactions of organic chemicals at the water surface or in the top layer of clear water. The presence of particles in a waterbody reduces light intensity through light scattering as does dissolved organic matter through light absorption. Photodegradation contributes significantly to removing oil spills and appears to favour the degradation of longer chain alkanes compared to the preferential attack of linear and small alkanes by biodegradation (Garrett et al., 1998). Cycloalkanes and aromatic hydrocarbons are also removed by photodegradation (D'Auria et al., 2009). There is comparatively little known about the role photodegradation of other organic pollutants in the marine environment although there is, for example, evidence that triclosan is removed by photolysis in the German Bight area of the North Sea (Xie et al., 2008). In the soil environment, there is some evidence that photodegradation may contribute to the removal of a variety of organic chemicals such as pesticides and chemicals present in sewage sludge that is used as a soil amendment but the significance of this process is unclear. Similarly, chemicals that have accumulated in ice, for example as a result of long range transport to polar regions, also seem to be susceptible to photodegradation. Some examples of photodegradation reactions are shown in Figure 4.

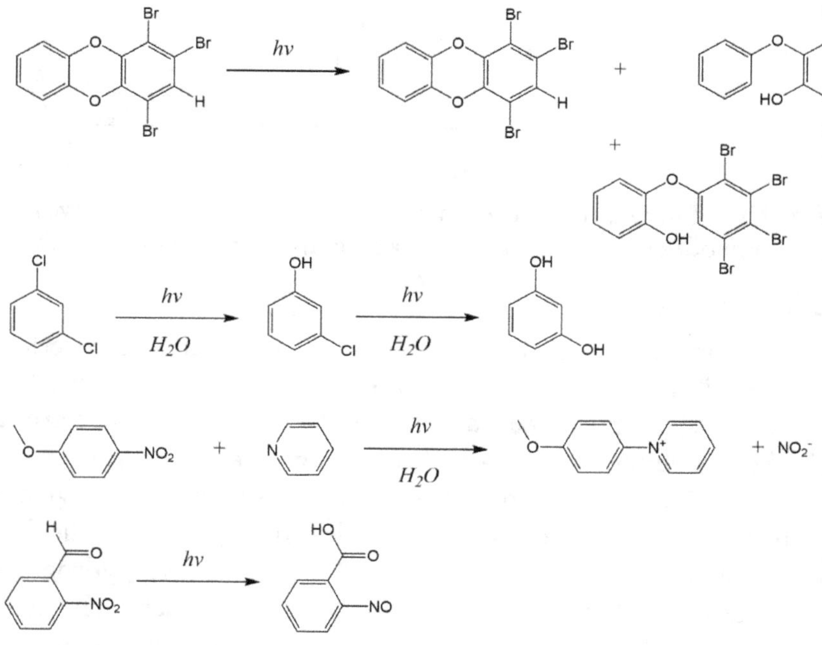

Figure 4. Some examples of photodegradation reactions (redrawn after Van Leeuwen and Vermeire, 2007). (Steven Droge 2019)

An important category of photochemical reactions are indirect reactions in which organic chemicals react with photochemically produced radicals, in particular with reactive oxygen species such as OH radicals, ozone and singlet oxygen. These reactive species are present at very low concentrations but are so reactive that under certain conditions they can contribute significantly to the removal of organic chemicals. Products of these reactions are a variety of oxidized derivatives which are themselves radicals and therefore react further. OH radicals are the most important of these photochemically produced species and can react with organic chemicals by removing hydrogen radicals, reacting with unsaturated bonds in alkenes, aromatics etc. to produce hydroxylated products. In water, natural organic matter absorbs light and can participate in indirect photodegradation reactions. Other constituents in surface water, such as nitrogen oxides and iron complexes may also be involved in indirect photodegradation reactions.

References

Schwarzenbach, R.P., Gschwend, P.M., Imboden, D.M. (2017). Environmental Organic Chemistry, Third Edition, Wiley, ISBN 978-1-118-76723-8

van Leeuwen, C.J., Vermeire, T.G. (2007). Risk Assessment of Chemicals: An Introduction (2nd ed.), Springer, ISBN 978-1-4020-6101-1

3.7.2. Biodegradation

Introduction

Biodegradation and biotransformation both refer to degradation reactions that are catalyzed by enzymes. In general, biodegradation is usually used to describe the degradation carried out by microorganisms and biotransformation often refers to reactions that follow the uptake of chemicals by higher organisms. This distinction is important and arises from the role that bacteria and other microorganisms play in natural biogeochemical cycles. As a result, microorganisms have the capacity to degrade most (perhaps all) naturally occurring organic chemicals in organic matter and convert them to inorganic end products. These reactions supply the microorganisms with the nutrients and energy they need to grow. This broad degradative capacity means that they are able to degrade many anthropogenic chemicals and potentially convert them to inorganic end products, a process referred to as mineralisation.

Although higher organisms are also able to degrade (metabolise) many anthropogenic chemicals, these chemicals are not taken up as source of nutrients and energy. Many anthropogenic chemicals can disturb cell functioning processes, and the biotransformation process has been proposed as a detoxification mechanism. Undesirable chemicals that may accumulate to potentially harmful levels are converted to products that are more rapidly excreted. In most cases, a polar and/or ionizable unit is attached to the chemical in one or two steps, making the compound more soluble in blood and more readily removed via the kidneys to the urine. This also renders most hazardous chemicals less toxic than the original chemical. Such biotransformation steps always costs energy (ATP, or through the use of e.g. NADH or NADPH in the enzymatic reactions) from the organism. Biotransformation is sometimes also used to describe degradation by microorganisms when this is limited to a conversion of a chemical into a new product.

Biodegradation is for many organic contaminants the major process that removes them from the environment. Measuring the rates of biodegradation therefore is a prominent aspect of chemical risk assessment. Internationally recognized standardised protocols have been developed to measure biodegradation rates of chemicals. Well know examples of these are the OCED Guidelines. These guidelines include screening tests designed to identify chemicals can be regarded as readily (i.e. rapidly) biodegradable as well as more complex tests to measure biodegradation rates of chemicals that degrade slowly in a variety of simulated environments. For more complex mechanistic studies, microorganisms able to degrade specific chemicals are isolated from environmental samples and cultivated in laboratory systems.

In principle, biodegradation of a chemical can be determined by either following the concentration of the chemical during the test or by following the conversion to end products (in most cases by either measuring oxygen consumption or CO2 production). Although measuring the concentration gives the most directly relevant information on a chemical, it requires the availability or development of analytical methods which is not always within the capability of routine testing laboratories. Measuring the conversion to CO2 is comparatively straightforward but the production of CO2 from other chemicals present in the test system (such as soil or dissolved organic matter) should be accounted for. This can be done by using 14C-labelled chemicals in the tests but not all laboratories have facilities for this. The main advantage of this approach is that demonstration of quantitative conversion of a chemical to CO2 etc. means that there is no concern about the accumulation of potentially toxic metabolites.

Since it is an enzymatically catalysed process, the rates of biodegradation should be modelled using the Michaelis Menten kinetics, or Monod kinetics if growth of the microorganisms is taken into account. In practice, however, first order kinetics are often used to model biodegradation in the absence of significant growth of the degrading microorganisms. This is more convenient that using Michaelis Menten kinetics but there is some justification for this simplification since the concentrations of chemicals in the environment are in general much lower than the half saturation concentrations of the degrading enzymes.

Table 1. Influence of molecular structure on the biodegradability of chemicals in the aerobic environment.

Type of compounds or substituents	More biodegradable	Less biodegradable
hydrocarbons	linear alkanes < C12	linear alkanes > C12
	alkanes with not too high molecular weight	high molecular weight alkanes
	linear chain	branched chain
	-C-C-C-	-C-O-C-

	aliphatic	aromatic
aliphatic chlorine	Cl more than 6 carbons from terminal C	Cl at less than 6 carbons from terminal C
Substituents to an aromatic ring	-OH	-F
	-CO2H	-Cl
	-NH2	-NO2
	-OCH3	-CF3

Whether expressed as terms of first order kinetics or Michaelis Menten parameters, rates of biodegradation vary widely for different chemicals showing that chemical structure has a large impact on biodegradation. Large variations in biodegradation rates are however often observed for the same chemical in different experimental systems. This shows that environmental properties and conditions also play a key role in determining removal by biodegradation and it is often almost impossible to distinguish the effects of chemical properties from those of environmental properties. In other words, there is no such thing as an intrinsic biodegradation rate of a chemical.

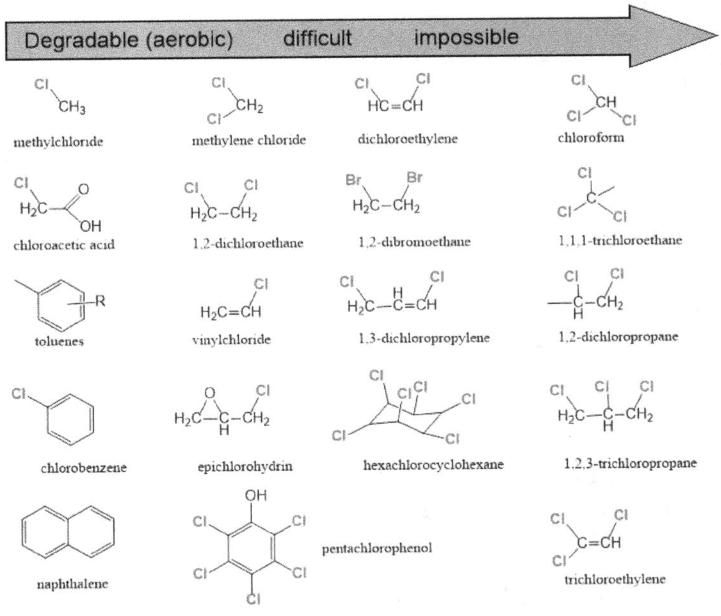

Figure 1. Effect of chlorination on (aerobic) biodegradation rates. Adapted from Janssen et al. (2005) by Steven Droge.

Nevertheless, we can derive some generic relationships between the structure and biodegradability of chemicals, as listed in Table 1. Examples are that branched hydrocarbon structures are degraded more slowly than linear hydrocarbon structures, and cyclic and in particular aromatic chemicals are degraded more slowly than aliphatic (non-aromatic) chemicals. Substituents and functional groups also have a major impact on biodegradability with halogens and other electron withdrawing substituents having strongly negative effects. It is therefore no surprise than the list of persistent organic pollutants is dominated by organohalogen compounds and in particular those with aromatic or alicyclic structures.

It should be recognized that biodegradation rates have often been observed to change over time. Long term exposure of microbial communities to new chemicals has often been observed to lead to increasing biodegradation rates. This phenomenon is called adaptation or acclimation and is often the case following repeated application of a pesticide at the same location. An example is shown for atrazine in Figure 2 where degradation rates increase following longer exposure to the pesticide.

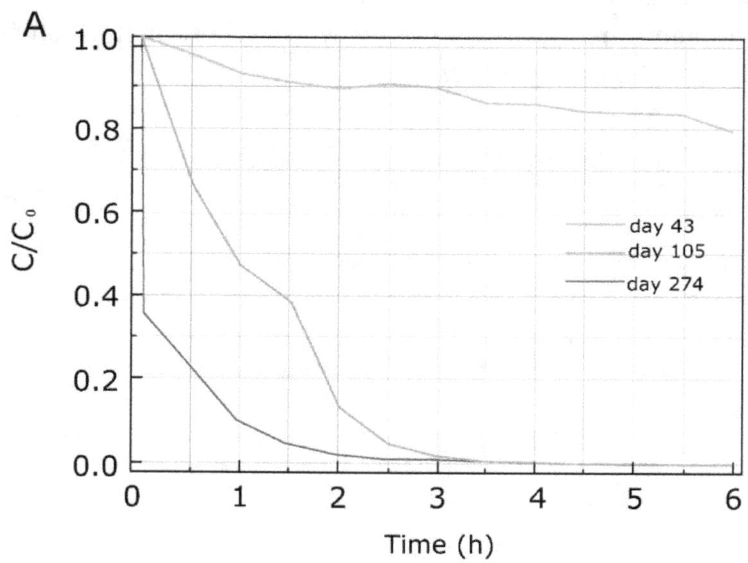

Figure 2. Comparison of the atrazine removal rates with (days 43 and 105) or without the addition of carbon and nitrogen sources (day 274). Redrawn from Zhou et al. (2017) by Wilma IJzerman.

Another recent example is the differences in biodegradation rates of the builder L-GLDA (tetrasodium glutamate diacetate) by activated sludge from different waste water treatment plants in the USA. Sludge from regions where L-GLDA was not or only recently on the market required long lag time before degradation started whereas sludge from regions where L-GLDA –containing products had been available for several months required shorted lag phases.

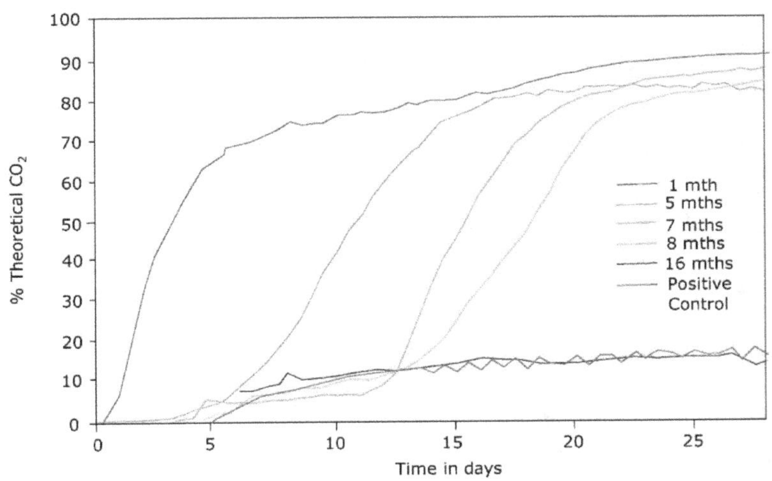

Figure 3. Biodegradation as a function of time following initial shipment of L-GLDA-containing products. Redrawn from Itrich et al. (2015) by Wilma Ijzerman.

Adaptation can results from i) shifts in composition or abundances of species in a bacterial community, ii) mutations within single populations, iii) horizontal transfer of DNA or iv) genetic recombination events, or combinations of these.

Biodegradation Reactions and Pathways

Biodegradation of chemicals that we regard as pollutants takes place when these chemicals are incorporated into the metabolism of microorganisms. The reactions involved in biodegradation are therefore similar to those involved in common metabolic reactions, such as hydrolyses, oxidations and reductions. Since the conversion of an organic chemical to CO_2 is an overall oxidation reaction, oxidation reactions involving molecular oxygen are probably the most important reactions. These reactions with oxygen are often the first but essential step in degradation and can be regarded as activation step converting relatively stable molecules to more reactive intermediates.

This is particularly important for aromatic chemicals since oxygenation is required to make aromatic rings susceptible to ring cleavage and further degradation. These reactions are catalysed by enzymes called oxygenases of which there are broadly speaking two classes. Monoxygenases are enzymes catalysing reactions in which one oxygen atom of O_2 reacts with an organic molecule to produce a hydroxylated product. Examples of such enzymes are the cytochrome P450 family and are present in all organisms. These enzymes are for example involved in the oxidation of alkanes to carboxylic acids as part of the "beta-oxidation" pathway, which shortens linear alkanoic acids in steps of C_2-units, as shown in Figure 4.

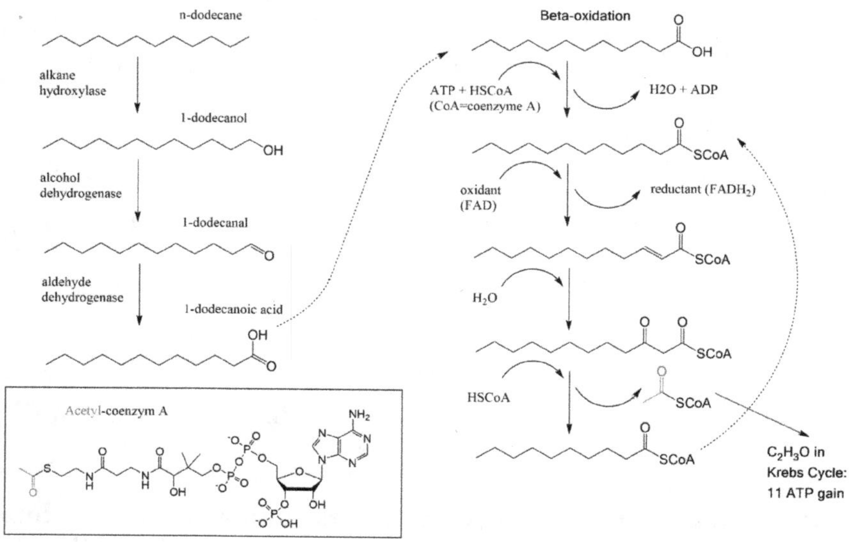

Figure 4. Typical oxidation steps of an alkane to an alkanoic acid, and subsequent beta-oxidation pathway from dodecanoic acid to decanoic acid, involving Coenzym-A. Redrawn from Schwarzenbach et al. (2003) by Steven Droge.

Dioxygenases are enzymes catalysing reactions in which both oxygen atoms of O_2 react with organic chemicals and appear to be unique to microorganisms such as bacteria. Examples of these reactions are shown for benzene in Figure 5. Similar reactions are involved in the degradation of more complex aromatic chemicals such as PAHs and halogenated aromatics.

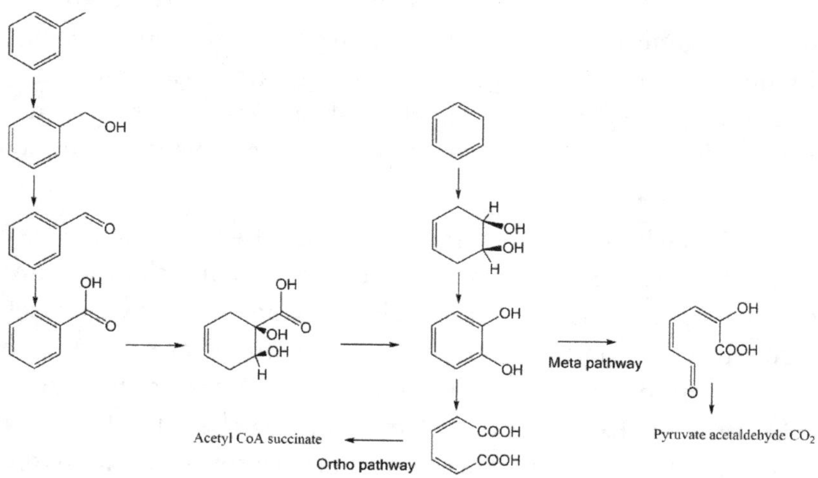

Figure 5. Examples of bacterial dioxygenation reactions and ring cleavage of toluene and benzene. Redrawn from Van Leeuwen and Vermeire (2007) by Steven Droge.

The absence of oxygen in anaerobic environments (sediments and groundwater) does not preclude oxidation of organic chemicals. Other oxidants present (nitrate, sulphate, Fe(III) etc) may be present in sufficiently high concentrations to act as oxidants and terminal electron acceptors supporting microbial growth. In the absence of oxygen, activation relies on other reactions, the most important reactions seem to be carboxylation or addition of fumarate. Figure 6 shows an example of the degradation of naphthalene to CO2 in sediment microcosms under sulphate-reducing conditions.

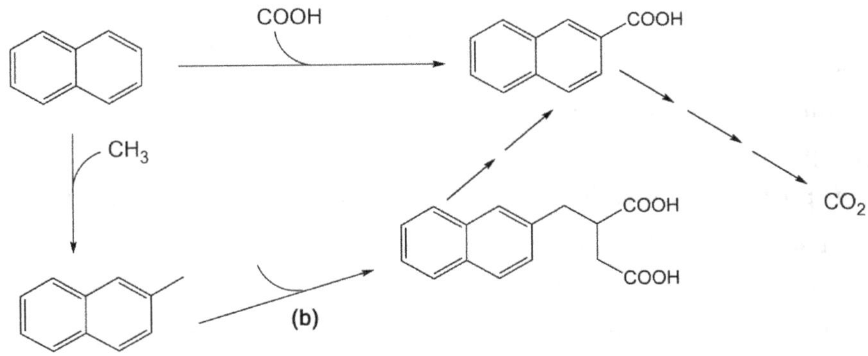

Figure 6. Proposed anaerobic oxidation pathway of naphthalene (redrawn from Kleeman and Merckenstock 2017). This involves two initial reaction mechanisms: carboxylation to naphthoic acid, methylation to 2-methylnaphthalene. Addition of fumarate (process b) could follow methylation and produce another way of transforming to 2-naphthoic acid. In both cases, 2-naphthoic acid is oxidized to CO_2.

Other important reactions in anaerobic ecosystems (sediments and groundwater plumes) are reductions. This affects functional groups, for example reduction of acids to aldehydes to alcohols, nitro groups to amino groups and, particularly important, substitution of halogens by hydrogen. The latter reactions can contribute to the conversion of highly chlorinated chemicals, that are resistant to oxidative biodegradation, to less chlorinated products which are more amenable to aerobic biodegradation. Many examples of these reductive dehalogenation reactions have been shown to occur in, for example, tetrachloroethene-contaminated groundwater (e.g. from dry-cleaning processes) and PCB-contaminated sediment. These reactions are exothermic under anaerobic conditions and some microorganisms are able to harvest this energy to support their growth. This can be considered to be a form of respiration based on dechlorination and is sometimes referred to as chlororespiration.

As is the case for abiotic degradation, hydrolyses are also important reactions in biodegradation pathways, particularly for chemicals that are derivatives of organic acids, such as carbamate, ester and organophosphate pesticides where hydrolyses are often the first step in their biodegradation. These reactions are similar to those described in the section on Chemical degradation.

References

Itrich, N.R., McDonough, K.M., van Ginkel, C.G., Bisinger, E.C., LePage, J.N., Schaefer, E.C., Menzies, J.Z., Casteel, K.D., Federle, T.W. (2015). Widespread microbial adaptation to L-glutamate-N,N,-diacetate (L-GLDA) following its market introduction in a consumer cleaning product. Environmental Science & Technology 49, 13314-13321.

Janssen, D. B., Dinkla, I. J. T., Poelarends, G. J., Terpstra, P. (2005). Bacterial degradation of xenobiotic compounds: evolution and distribution of novel enzyme activities, Environmental Microbiology 7, 1868-1882.

Kleemann, R., Meckenstock, R.U. (2017). Anaerobic naphthalene degradation by Gram-positive, iron-reducing bacteria. FEMS Microbial Ecology 78, 488-496.

Schwarzenbach, R.P., Gschwend, P.M., Imboden, D.M. (2017). Environmental Organic Chemistry, Third Edition, Wiley, ISBN 978-1-118-76723-8

Van Leeuwen, C., Vermeire, T.G. (2007). Risk Assessment of Chemicals: An Introduction (2nd ed.), Springer, ISBN 978-1-4020-6101-1

Zhou, Q., Chen, L. C., Wang, Z., Wang, J., Ni, S., Qiu, J., Liu, X., Zhang, X., Chen, X. (2017). Fast atrazine degradation by the mixed cultures enriched from activated sludge and analysis of their microbial community succession. Environmental Science & Pollution Research 24, 22152-22157.

3.7.3. Degradation Test Methods

Introduction

Many experimental approaches are possible to measure the environmental degradation of chemicals, ranging from highly controlled laboratory experiments to environmental monitoring studies. While each of these approaches has its advantages and disadvantages, a standardised and relatively straightforward set of protocols has clear advantages such as suitability for a wide range of laboratories, broad scientific and regulatory acceptance and comparability for different chemicals.

The system of OECD test guidelines (see links in the reference list of this chapter) is the most important set of standardised protocols although other test systems may be used in other regulatory contexts. As well as tests covering environmental fate processes, they also cover physical-chemical properties, bioaccumulation, toxicity etc. These guidelines have been developed in an international context and are adopted officially after extensive validation and testing in different laboratories.

This ensures their wide acceptance and application in different regulatory contexts for chemical hazard and risk assessment.

Chemical Degradation Tests

The OECD Guidelines include only two tests specific for chemical degradation. This might seem surprising but it should not be forgotten that chemical degradation could also contribute to the removal observed in biodegradability tests. The OECD Guidelines for chemical degradation are OECD Test 111: Hydrolysis as a Function of pH (OECD 2004) and OECD Test 316: Phototransformation of Chemicals in Water – Direct Photolysis (OECD 2008). If desired, sterilised controls may also be used to determine the contribution of chemical degradation in biodegradability tests.

OECD Test 111 measures hydrolytic transformations of chemicals in aquatic systems at pH values normally found in the environment (pH 4 – 9). Sterile aqueous buffer solutions of different pH values (pH 4, 7 and 9) containing radio-labelled or unlabelled test substance (below saturation) are incubated in the dark at constant temperature and analysed after appropriate time intervals for the test substance and for hydrolysis products. The preliminary test is carried out for 5 days at 50°C and pH 4.0, 7.0 and 9.0, this is known as a first tier test. Further second tier tests study the hydrolysis of unstable substances and the identification of hydrolysis products and may extend for 30 days.

OECD Test 316 measures direct photolysis rate constants using xenon arc lamp capable of simulating natural sunlight in the 290 to 800 nm or natural sunlight, and extrapolated to natural water. If estimated losses are superior or equal to 20%, the transformation pathway and the identities, concentrations, and rate of formation and decline of major transformation products are identified.

Biodegradability Tests

Biodegradation is in general considered to be the most important removal process for organic chemicals in the environment and it is therefore no surprise that biodegradability testing plays a key role in the assessing the environmental fate and subsequent exposure risks of chemicals. Biodegradation is an extensively researched area but data from standardised tests are favoured for regulatory purposes as they are assumed to yield reproducible and comparable data. Standardised tests have been developed internationally, most importantly under the auspices of the OECD and are part of the wider range of tests to measure physical-chemical, environmental and toxicological properties of chemicals. An overview of these biodegradability tests is given in Table 1.

The way that biodegradability testing is implemented can vary in detail depending on the regulatory context but in general it is based on a tiered approach with all chemicals being subjected to screening tests to identify chemicals that can be considered to be readily biodegradable and therefore removed rapidly from wastewater treatment plants (WWTPs) and the environment in general. These tests were originally developed for surfactants and often use activated sludge from WWTPs as a source of microorganisms since biodegradation during wastewater treatment is a major conduit of chemical emissions to the environment. The so-called ready biodegradability tests are designed to be stringent with low bacterial concentrations and the test chemical as the only potential source of carbon and energy at high concentrations. The assumption is that chemicals that show rapid biodegradation under these unfavourable conditions will always be degraded rapidly under environmental conditions. Biodegradation is determined as conversion to CO_2 (mineralisation), either by directly measuring CO_2 produced, or the consumption of oxygen, or removal of dissolved organic carbon, as this is the most desirable outcome of biodegradation. The results that have to be achieved for a chemical to be considered readily biodegradable vary slightly depending on the test, but as an example in the OECD 301D test (OECD 2014), the consumption of oxygen should reach 70% of that theoretically required for complete mineralisation within 28 days.

Table 1. The OECD biodegradability tests

OECD TEST GUIDELINE	PARAMETER MEASURED	REFERENCE
Ready biodegradability tests		
301A: DOC Die-away test	DOC	OECD 1992a
301B: CO2 evolution test	CO_2	OECD 1992a
301C: Modified MITI(I) test	O_2	OECD 1992a
301D: Closed bottle test	O_2	OECD 1992a
301E: Modified OECD screening test	DOC	OECD 1992a
301F: Manometric respirometry test	O_2	OECD 1992a
306: Biodegradability in seawater	DOC	OECD 1992c
310: Test No. 310: Ready Biodegradability - CO2 in sealed vessels (Headspace Test).	CO_2	OECD 2014
Inherent biodegradability tests		
302A: Modified Semi-continuous Activated Sludge (SCAS) test	DOC	OECD 1981b
302B: Zahn-Wellens test	DOC	OECD 1992b
302C: Modified MITI(II) test	O_2	OECD 2009
Simulation tests		
303A: Activated sludge units	DOC	OECD 2001
303B: Biofilms	DOC	OECD 2001
304A: Inherent biodegradability in soil	$^{14}CO_2$	OECD 1981a

OECD TEST GUIDELINE	PARAMETER MEASURED	REFERENCE
307: Aerobic and anaerobic transformation in soil	$14CO_2/CO_2$	OECD 2002a
308: Aerobic and anaerobic transformation in aquatic sediment systems	$14CO_2/CO_2$	OECD 2002b
309: Aerobic mineralization in surface water	$14CO_2/CO_2$	OECD 2004b
311: Anaerobic biodegradability of organic compounds in digested sludge: by measurement of gas production	CO_2 and CH_4	OECD 2006
314: Simulation tests to assess the biodegradability of chemicals discharged in wastewater	Concentration of chemical, $14CO_2/CO_2$	OECD 2008a

These test systems are widely applied for regulatory purposes but they do have a number of issues. These include the fact that there are practical difficulties when applied to volatile or poorly soluble chemicals, but probably the most important is that for some chemicals the results can be highly variable. This is usually attributed to the source of the microorganisms used to inoculate the system. For many chemicals, there is a wide variability in how quickly they are degraded by activated sludge from different WWTPs. This is probably the result of different exposure concentrations and exposure periods to the chemicals, and may also be caused by dependence on the ability of small populations of degrading microorganisms, which may not always be included in the sludge samples used in the tests. These issues are not dealt with in any systematic way in biodegradability testing. It has been suggested that a preliminary period of exposure to the chemicals to be tested would allow sludge to adapt to the chemicals and may yield more reproducible test results. Further suggestions include using a higher, more environmentally relevant, concentration of activated sludge as the inoculum.

Failure to comply with the pass criteria in ready biodegradability tests does not necessarily mean that the chemical is persistent in the environment since it is possible that slow biodegradation may occur. These chemicals may therefore be tested further in higher tier tests, for what is referred to as inherent biodegradability in tests performed under more favourable conditions or in simulation tests representing specific compartments, to determine whether biodegradation may contribute significantly to their removal. These tests are also standardised (see Table 1). Simulation tests are designed to represent environmental conditions in specific compartments, such as redox potential, pH, temperature, microbial community, concentration of test substance and occurrence and concentration of other substrates.

The criteria used in classifying the biodegradability of chemicals depend on the regulatory context. Biodegradability tests can be used for different purposes:

in the EU this includes 3 distinct purposes; classification and labelling, hazard/ persistent assessment, and environmental risk assessment Recently regulatory emphasis has shifted to identifying hazardous chemicals, and therefore those chemicals that are less biodegradable and likely to persist in the environment. Examples for the classification as PBT (persistent, bioaccumulative and toxic) or vPvB (very persistent and very bioaccumulative) chemicals are shown in Table 2. As well as the results of standardised tests, other data such as the results of environmental monitoring data or studies on the microbiology of biodegradation can also be taken into account in evaluations of environmental degradation in a so-called weight of evidence approach.

Table 2. Criteria used to classify chemicals as PBT or vPvB (van Leeuwen & Vermeire 2007)

Property	PBT Criteria	vPvB Criteria
Persistence	$T_{1/2}$ >60 days in marine water, or $T_{1/2}$ >40 days in fresh/estuarine water, or $T_{1/2}$ >180 days in marine sediment, or $T_{1/2}$ >120 days in fresh/estuarine sediment, or $T_{1/2}$ >120 days in soil.	$T_{1/2}$ >60 days in marine, fresh or estuarine water, or $T_{1/2}$ >180 days in marine, fresh or estuarine sediment, or $T_{1/2}$ >180 days in soil
Bioaccumulation	BCF > 2000 L/kg	BCF > 5000 L/kg
Toxicity	• NOEC < 0.01 mg/L for marine or freshwater organisms, or • substance is classified as carcinogenic, mutagenic, or toxic for reproduction, or • there is other evidence of chronic toxicity, according to Directive 67/548/EEC	

The results of biodegradability tests are sometimes also used to derive input data for environmental fate models (see section on Multicompartment modeling). It is however not always straightforward to transfer data measured in what is sometimes a multi-compartment test system into degradation rates in individual compartments as other processes (e.g. partitioning) need to be taken into account.

References

OECD, 1981a. OECD Guidelines for the Testing of Chemicals. Test No. 304A: Inherent Biodegradability in Soil.

OECD, 1981b. OECD Guidelines for the Testing of Chemicals. Test No. 302A: Inherent Biodegradability: Modified SCAS Test.

OECD, 1992a. OECD Guidelines for the Testing of Chemicals. Test No. 301: Ready Biodegradability.

OECD, 1992b. OECD Guidelines for the Testing of Chemicals. Test No. 302B: Inherent Biodegradability: Zahn-Wellens/ EVPA Test.

OECD, 1992c. OECD Guidelines for the Testing of Chemicals. Test No. 306: Biodegradability in Seawater.

OECD, 2001. OECD Guidelines for the Testing of Chemicals. Test No. 303: Simulation Test - Aerobic Sewage Treatment - A: Activated Sludge Units; B: Biofilms.

OECD, 2002a. OECD Guidelines for the Testing of Chemicals. Test No. 307: Aerobic and Anaerobic Transformation in Soil.

OECD, 2002b. OECD Guidelines for the Testing of Chemicals. Test No. 308: Aerobic and Anaerobic Transformation in Aquatic Sediment Systems.

OECD, 2004a. OECD Guidelines for the Testing of Chemicals. Test No. 111: Hydrolysis as a Function of pH.

OECD, 2004b. OECD Guidelines for the Testing of Chemicals. Test No. 309: Aerobic Mineralisation in Surface Water – Simulation Biodegradation Test

OECD, 2006. OECD Guidelines for the Testing of Chemicals. Test No. 311: Anaerobic Biodegradability of Organic Compounds in Digested Sludge: by Measurement of Gas Production.

OECD, 2008a. OECD Guidelines for the Testing of Chemicals. Test No. 314: Simulation Tests to Assess the Biodegradability of Chemicals Discharged in Wastewater

OECD, 2008b. OECD Guidelines for the Testing of Chemicals. Test No. 316: Phototransformation of Chemicals in Water – Direct Photolysis.

OECD, 2009. OECD Guidelines for the Testing of Chemicals. Test No. 302C: Inherent Biodegradability: Modified MITI Test (II).

OECD, 2014. OECD Guidelines for the Testing of Chemicals. Test No. 310: Ready Biodegradability - CO2 in sealed vessels (Headspace Test).

Van Leeuwen, C.J., Vermeire, T.G. (2007). Risk Assessment of Chemicals: An Introduction (2nd ed.), Springer, ISBN 978-1-4020-6101-1

3.8. MODELLING EXPOSURE

3.8.1. Multicompartment Modeling

The Mass Balance Equation

Multicompartment (or multimedia) mass balance modeling starts from the universal conservation principle, formulated as balance equation. The governing principle is that the rate of change (of any entity, in any system) equals the difference between the sum of all inputs (of that entity) to the system and the sum of all outputs from it. Environmental modelers use the balance equation to predict exposure concentrations of chemicals in the environment by deduction from knowledge of the rates of input- and output processes, which can be understood easiest from considering the mass balance equation for one single environmental compartment (Figure 1):

$$\frac{dm^{i,j}}{dt} = \sum^{j} input^{i,j} - \sum^{j} output^{i,j} \qquad \text{(eq. 1)}$$

where dmi,j/dt represents the change of mass of chemical i in compartment j (kg) over time (s), and $input^{i,j}$ and $output^{i,j}$ denote the rates of input and output of chemical to and from compartment j, respectively.

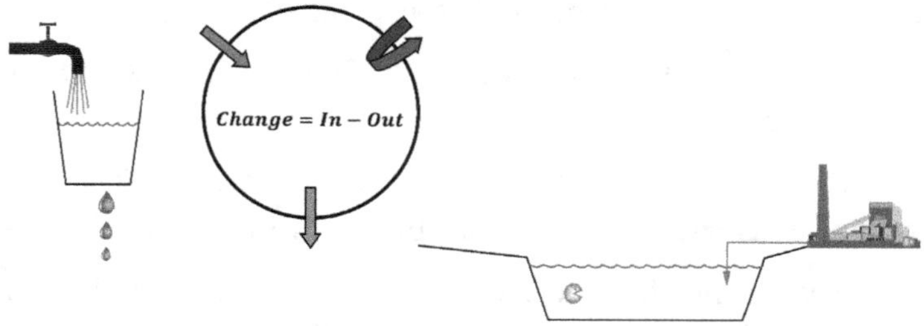

Figure 1. The mass of a chemical in a lake is like the mass of water in a leaking bucket: both can be described with the universal (mass) balance equation, which says differences in inputs and outputs make amounts held up in systems change:

$$\frac{dm}{dt} = \sum inputs - \sum outputs$$

One Compartment Model

In multimedia mass balance modeling, mass balance equations (of the type shown in equation 1) are formulated for each environmental compartment. Outflows of chemical from the compartments are often proportional to the amounts of chemical present in the compartments, while external inputs (emissions) may often be assumed constant. In such cases, i.e. when first-order kinetics apply (see section 3.3 on Environmental fate of chemicals), mass balance equations take the form of equation 1 in section 3.3. For one compartment (e.g. a lake, as in Figure 1) only:

$$\frac{dm}{dt} = I - km \qquad \text{(eq. 2)}$$

In which dm/dt (kg.s^{-1}) is the rate of change of the mass (kg) of chemical in the lake, I (kg.s^{-1}) is the (constant) emission rate, and the product $k.m$ (kg.s^{-1}) denotes the first-order loss rate of the chemical from the lake. It is obvious that eventually a steady state must develop, in which the mass of chemical in the lake reaches a predictable maximum

$$\frac{dm}{dt} = I - k\,m = 0 \rightarrow m_{\infty} = \frac{I}{k} \qquad \text{(eq. 2a)}$$

A very intuitive result: mass (thus concentration) at steady state is proportional to the rate of emission (twice the emission yields twice the mass or concentration); steady-state mass is inversely proportional to the rate (constant) of degradation (more readily degrading chemicals reach lower steady-state masses). It can be demonstrated mathematically (not here) that the general (transient) solution of equation 2 exists and can be found (Figure 2):

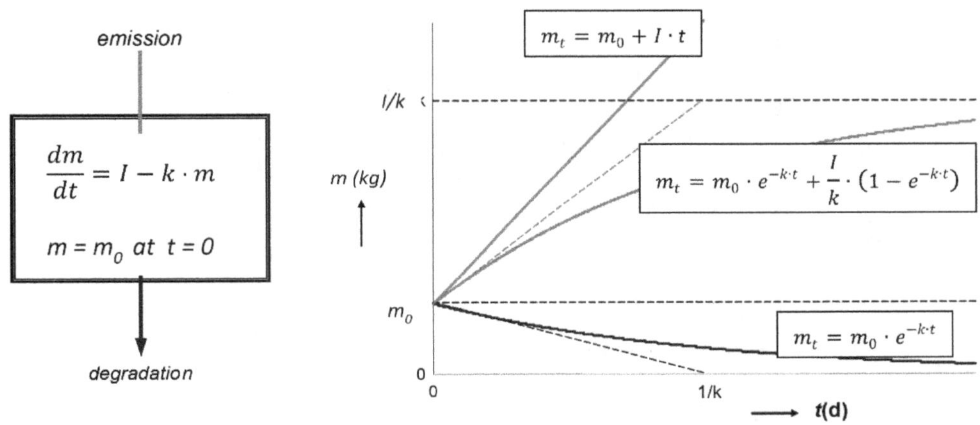

Figure 2. For one compartment only, in case loss processes obey first-order kinetics and emissions are constant (i.e. not varying with time), the mass of chemical is expected to increase exponentially from its initial mass m0 to its steady-state level I , which will be reached at ininite time t∞. After Van de Meent et al. (2011).

$$m_t = m_0\, e^{-k\,t} + \frac{I}{k}(1 - e^{-k\,t}), with \frac{I}{k} = m_{\infty} \qquad \text{(eq. 3)}$$

When the input rate (emission) is constant, i.e. that it does not vary with time, and is independent of the mass of chemical present, the mass of chemical in the systems is expected to increase exponentially, from its initial value at t0, to a steady level at t∞. According to equation 3, a final mass level equal to $\frac{I}{k}$ is to be expected.

Multi-compartment Model

The prefix 'multi' indicates that generally (many) more than one environmental compartment is considered. The Unit World (see below) contains air, water, biota, sediment and soil; more advanced global modelling systems may use hundreds of compartments. The case of three compartments (typically one air, one water, one soil) is schematically worked out in Figure 3.

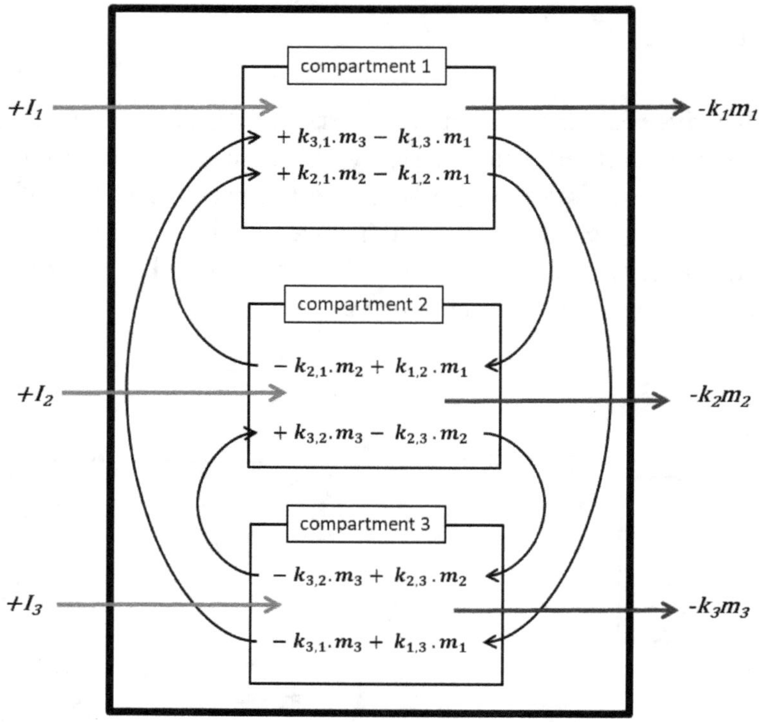

$$\frac{dm_1}{dt} = I_1 - (k_1 + k_{1,2} + k_{1,3})m_1 + k_{2,1}\, m_2 + k_{3,1}\, m_3$$

$$\frac{dm_2}{dt} = I_2 + k_{1,2}\, m_1 - (k_2 + k_{2,1} + k_{2,3})m_2 + k_{3,2}\, m_3$$

$$\frac{dm_3}{dt} = I + k_{1,3}\, m_1 + k_{2,3}\, m_2 - (k_3 + k_{3,1} + k_{3,2}).m_3$$

Figure 3. Three-compartment mass balance model. Arrows represent mass flows of chemical to and from compartments. Losses from source compartments (negative signs) become gains to receiving compartments (positive sign). The model consists of three differential mass balance equations, with (nine) known rate constants ki,j (for flows out of source compartments i, into receiving compartments j, in s-1), and (three) unknown masses mi (kg). From Van de Meent et al. (2011), with permission.

Each compartment can receive constant inputs (emissions, imports), and chemical can be exported from each compartment by degradation or advective outflow, as in the one-compartment model. In addition, chemical can be transported between compartments (simultaneous import-export). All mass flows are characterized by (pseudo) first-order rate constants (see section 3.3 on Environmental fate processes). The three mass balance equations eventually balance to zero at infinite time:

$$balance_1 = 0 = I_1 - (k_1 + k_{1,2} + k_{1,3})m_1^* + k_{2,1}\, m_2^* + k_{3,1}\, m_3^*$$
$$balance_2 = 0 = I_2 + k_{1,2}\, m_1^* - (k_2 + k_{2,1} + k_{2,3})m_2^* + k_{3,2}\, m_3^*$$
$$balance_3 = 0 = I_3 + k_{1,3}\, m_1^* + k_{2,3}\, m_2^* - (k_3 + k_{3,1} + k_{3,2})m_3^*$$
(eq. 4)

where the symbols m_i^* denote mass in compartments i at steady state. Sets of n linear equations with n unknowns can be solved algebraically, by manually manipulating equations 4, until clean expressions for each of the three mi values are obtained, which inevitably becomes tedious as soon as more than two mass balance equations are to be solved – this did not withhold one of Prof. Mackay's most famous PhD students from successfully solving a set of 14 equations! An easier way of solving sets of n linear equations with n unknowns is by means of linear algebra. Using linear algebraic vector-matrix calculus, the equations 4 can be rewritten into one linear-algebraic equation:

$$\frac{d\bar{m}}{dt} = 0 = \bar{I} + A\,\bar{m}$$
(eq. 5)

in which in which ^{mo} is the vector of masses in the three compartments, A is the model matrix of known rate constants and ^{eo} is the vector of known emission rates:

$$\bar{m} = \begin{bmatrix} m_1 \\ m_2 \\ m_3 \end{bmatrix},\ \bar{I} = \begin{bmatrix} I_1 \\ I_2 \\ I_3 \end{bmatrix},\ A = \begin{bmatrix} -(k_1 + k_{1,2} + k_{1,3}) & k_{2,1} & k_{3,1} \\ k_{1,2} & -(k_2 + k_{2,1} + k_{2,3}) & k_{3,2} \\ k_{1,3} & k_{2,3} & -(k_3 + k_{3,1} + k_{3,2} \end{bmatrix}$$

The solution of equation 5 is

$$\bar{m}^* = -A^{-1}\,\bar{I}$$

in which $m_\circ^*$ is the vector of masses at steady state and A-1 is the inverse of model matrix A. The linear algebraic method of solving linear mass balance equations is easily carried with spreadsheet software (such as MS Excel, LibreOffice Calc or Google Sheets), which contain built-in array functions for inverting matrices and multiplying them by vectors.

Unit World Modeling

In the late 1970s, pioneering environmental scientists at the USEPA Environmental Research Laboratory in Athens GA, recognized that the universal (mass) balance equation, applied to compartments of environmental media (air, water, biota, sediment, soil) could serve as a means to analyze and understand differences in environmental behavior and fate of chemicals. Their 'evaluative Unit World Modeling' (Baughman and Lassiter, 1978; Neely and Blau, 1985) was the start of what is now known as multimedia mass balance modeling. The Unit World concept was further developed and polished by Mackay and co-workers (Neely and Mackay, 1982; Mackay and Paterson, 1982; Mackay et al., 1985; Paterson and Mackay, 1985, 1989). In Unit World modeling, the environment is viewed of as a set of well-mixed chemical reactors, each representing one environmental medium (compartment), to and from which chemical flows, driven by 'departure from equilibrium' – this is chemical technology jargon for expressing the degree to which thermodynamic equilibrium properties such as 'chemical potential' or 'fugacity' differ (Figure 4). Mackay and co-workers used fugacity in mass balance modeling as the central state variable. Soon after publication of this 'fugacity approach' (Mackay, 1991), the term 'fugacity model' became widely used to name all models of the 'Mackay-type', which applied 'Unit World mass balance modeling', even though most of these models kept using the more traditional chemical mass as a state variable.

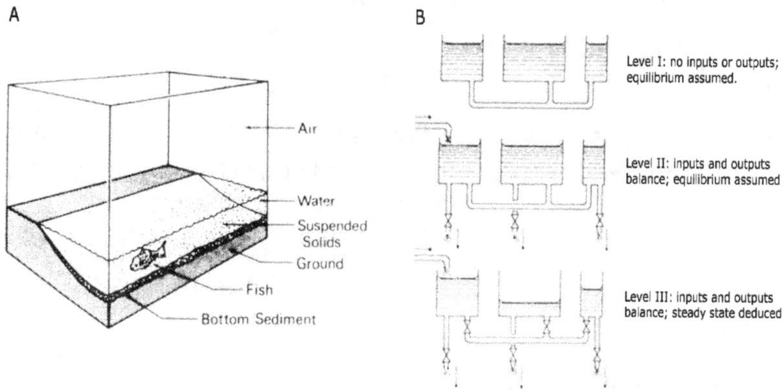

Figure 4. Unit World mass balance modeling as described by Mackay and co-workers. The environment is viewed of as a set of well-mixed chemical reactors (A), for which mass balance equations are formulated. Chemical flows from one environmental compartment to another, driven by 'departure from equilibrium', until a steady (= not changing) state has been reached. This may be understood by regarding the hydraulic equivalent of chemical mass flow (B). Figure redrawn after Mackay (1991).

Complexity Levels

While conceptually simple (environmental fate is like a leaking bucket, in the sense that its steady-state water height is predictable from first-order kinetics), the dynamic character of mass balance modeling is often not so intuitive. The abstract mathematical perspective may best suit explain mass balance modeling, but this may not be practical for all students. In his book about multimedia mass balance modeling, Mackay chose to teach his students the intuitive approach, by means of his famous water tank analogy (Figure 4B).

According to this intuitive approach, mass balance modeling can be done at levels of increasing complexity, where the lowest, simplest, level that serves the purpose should be regarded as the most suitable. The least complex is level I assuming no input and output. A chemical can freely (i.e. without restriction) flow from one environmental compartment to another, until it reaches its state of lowest energy: the state of thermodynamic equilibrium. In this state, the chemical has equal chemical potential and fugacity in all environmental media. The system is at rest; in the hydraulic analogy, water has equal levels in all tanks. This is the lowest level of model complexity, because this model only requires knowledge of a few thermodynamic equilibrium constants, which can be reasoned from basic physical substance properties.

The more complex modeling level III describes an environment in which flow of chemical between compartments experiences flow resistance, so that a steady state of balance between outputs and inputs is reached only at the cost of permanent 'departure from equilibrium'. Degradation in all compartments and advective flows, e.g. rain fall or wind and water currents, are also considered. The steady state of level III is one in which fugacities of chemical in the compartments are unequal (no thermodynamic equilibrium); in the hydraulic analogy, water in the tanks rest at different heights. Naturally, solving modeling level III requires detailed knowledge of the inputs (into which compartment(s) is the chemical emitted?), the outputs (at what rates is the chemical degraded in the various compartments?) and the transfer resistances (how rapid or slow is the mass transfer between the various compartments?). Level III modelers are rewarded for this by obtaining more realistic model results.

The fourth complex level of multimedia mass balance modeling (level IV, not shown in Figure 4B) produces transient (time dependent) solutions. Model simulations start ($t = 0$) with zero chemical ($m = 0$; empty water tanks). Compartments (tanks), fill up gradually until the system comes to a steady state, in which generally one or more compartments depart from equilibrium, as in level III modeling. Level IV is the most realistic representation of environmental fate of chemicals, but requires most detailed knowledge of mass flows and mass transfer resistances. Moreover, time-varying states are least easy to interpret and not always most informative of chemical fate. The most important piece of information to be gained from level IV modeling is the indication of time to steady state: how long does it take to clear the environment from persistent chemicals that are no longer used?

Mackay describes an intermediate level of complexity (level II), in which outputs (degradation, advective outflows) balance inputs (as in level III), and chemical is allowed to freely flow between compartments (as in level I). A steady state develops in level II and there is thermodynamic equilibrium at all times. Modeling at level II does not require knowledge of mass transfer resistances (other than that resistances are negligible!), but degradation and outflow rates increase the model complexity compared to that of level I. In many situations, level II modeling yields surprisingly realistic results.

Use of Multimedia Mass Balance Models

Soon after publication of the first use of 'evaluative Unit World modeling' (Mackay and Paterson, 1982), specific applications of the 'Mackay approach' to multimedia mass balance modeling started to appear. The Mackay group published several models for the evaluation of chemicals in Canada, of which ChemCAN (Mackay et al., 1995) is known best. Even before ChemCAN, the Californian model CalTOX (Mckone, 1993) and the Dutch model SimpleBox (Van de Meent, 1993) came out, followed by publication of the model HAZCHEM by the European Centre for Ecotoxicology and Toxicology of Chemicals (ECETOC, 1994) and the German Umwelt Bundesamt's model ELPOS (Beyer and Matthies, 2002). Essentially, all these models serve the very same purpose as the original Unit World model, namely providing standardized modeling platforms for evaluating the possible environmental risks from societal use of chemical substances.

Multimedia mass balance models became essential tools in regulatory environmental decision making about chemical substances. In Europe, chemical substances can be registered for marketing under the REACH regulation only when it is demonstrated that the chemical can be used safely. Multimedia mass balance modeling with SimpleBox (Hollander et al., 2014) and SimpleTreat (Struijs et al, 2016) plays an important role in registration.

While early multimedia mass balance models all followed in the footsteps of Mackay's Unit World concept (taking the steady-state approach and using one compartment per environmental medium), later models became larger and spatially and temporally explicit, and were used for in-depth analysis of chemical fate.

In the late 1990s, Wania and co-workers developed a Global Distribution Model for Persistent Organic Pollutants (GloboPOP). They used their global multimedia mass balance model to explore the so-called cold condensation effect, by which they explained the occurrence of relatively large amounts of persistent organic chemicals in the Arctic, where no one had ever used them (Wania, 1999). Scheringer and co-workers used their CliMoChem model to investigate long-range transport of persistent chemicals into Alpine regions (Scheringer, 1996; Wegmann et al., 2005). MacLeod and co-workers (Toose et al., 2004) constructed a global multimedia mass balance model (BETR World) to study long-range, global transport of pollutants.

References

Baughman, G.L., Lassiter, R. (1978). Predictions of environmental pollutant concentrations. In: Estimating the Hazard of Chemical Substances to Aquatic Life. ASTM STP 657, pp. 35-54.

Beyer, A., Matthies, M. (2002). Criteria for Atmospheric Long-range Transport Potential and Persistence of Pesticides and Industrial Chemicals. Umweltbundesamt Berichte 7/2002, E. Schmidt-Verlag, Berlin. ISBN 3-503-06685-3.

ECETOC (1994). HAZCHEM, A mathematical Model for Use in Risk Assessment of Substances. European Centre for Ecotoxicology and Toxicology of Chemicals, Brussels.

Hollander, A., Schoorl, M., Van de Meent, D. (2016). SimpleBox 4.0: Improving the model, while keeping it simple... Chemosphere 148, 99-107.

Mackay, D. (1991). Multimedia Environmental Fate Models: The Fugacity Approach. Lewis Publishers, Chelsea, MI.

Mackay, D., Paterson, S. (1982). Calculating fugacity. Environmental Science and Technology 16, 274-278.

Mackay, D., Paterson, S., Cheung, B., Neely, W.B. (1985). Evaluating the environmental behaviour of chemicals with a level III fugacity model. Chemosphere 14, 335-374.

Mackay, D., Paterson, S., Tam, D.D., Di Guardo, A., Kane, D. (1995). ChemCAN: A regional Level III fugacity model for assessing chemical fate in Canada. Environmental Toxicology and Chemistry 15, 1638–1648.

McKone, T.E. (1993). CALTOX, A Multi-media Total-Exposure Model for Hazardous Waste sites. Lawrence Livermore National Laboratory. Livermore, CA.

Neely, W.B., Blau, G.E. (1985). Introduction to Exposure from Chemicals. In: Neely, W.B., Blau, G.E. (Eds). Environmental Exposure from Chemicals Volume I, CRC Press, Boca Raton, FL., pp 1-10.

Neely, W.B., Mackay, D. (1982). Evaluative model for estimating environmental fate. In: Modeling the Fate of Chemicals in the Aquatic Environment. Ann Arbor Science, Ann Arbor, MI, pp. 127-144.

Paterson, S. (1985). Equilibrium models for the initial integration of physical and chemical properties. In: Neely, W.B., Blau, G.E. (Eds). Environmental Exposure from Chemicals Volume I, CRC Press, Boca Raton, FL., pp 218-231.

Paterson, S., Mackay, D. (1989). A model illustrating the environmental fate, exposure and human uptake of persistent organic chemicals. Ecological Modelling 47, 85-114.

Scheringer, M. (1996). Persistence and spatial range as endpoints of an exposure-based assessment of organic chemicals. Environmental Science and Technology 30, 1652-1659.

Struijs,J., Van de Meent, D., Schowanek, D., Buchholz, H., Patoux, R., Wolf, T., Austin, T., Tolls, J., Van Leeuwen, K., Galay-Burgos, M. (2016). Adapting SimpleTreat for simulating behaviour of chemical substances during industrial sewage treatment. Chemosphere 159:619-627.

Toose, L., Woodfine, D.G., MacLeod, M., Mackay, D., Gouin, J. (2004). BETR-World: a geographically explicit model of chemical fate: application to transport of alpha-HCH to the Arctic Environmental Pollution 128, 223-40.

Van de Meent D. (1993). SimpleBox: A Generic Multi-media Fate Evaluation Model. National Institute for Public Health and the Environment. RIVM Report 672720 001. Bilthoven, NL.

Van de Meent, D., McKone, T.E., Parkerton, T., Matthies, M., Scheringer, M., Wania, F., Purdy, R., Bennett, D. (2000). Persistence and transport potential of chemicals in a multimedia environment. In: Klecka, g. et al. (Eds.) Evaluation of Persistence and Long-Range Transport Potential of Organic Chemicals in the Environment. SETAC Press, Pensacola FL, Chapter 5, pp. 169–204.

Van de Meent, D., Hollander, A., Peijnenburg, W., Breure, T. (2011). Fate and transport of contaminants. In: Sánches-Bayo, F., Van den Brink, P.J., Mann, R.M. (eds.),Ecological Impacts of Toxic Chemicals, Bentham Science Publishers, pp. 13-42.

Wania, F. (1999). On the origin of elevated levels of persistent chemicals in the environment. Environmental Science and Pollution Research 6, 11–19.

Wegmann, F., Scheringer, M., Hungerbühler, K. (2005). First investigations of mountainous cold condensation effects with the CliMoChem model. Ecotoxicology and Environmental Safety 63, 42-51.

3.8.2. Metal Speciation Models

Introduction

Speciation models allow users to calculate the speciation of a solution rather than to measure it in a chemical way or to assess it indirectly using bioassays (see section 3.5). As a rule, speciation models take total concentrations as input and calculate species concentrations.

Speciation models use thermodynamic data about chemical equilibria to calculate the speciation. This data, expressed in free energy or as equilibrium constants, can be found in the literature. The term 'constant' is slightly misleading as equilibrium constants depend on the temperature and ionic strength of the solution. The ionic strength is calculated from the concentrations (C) and charges (Z) of ions in solution using the equation:

$$1 = \frac{1}{2}\sum_{i=1}^{n} C_i Z_i^2$$

For many equilibria, no information is available to correct for temperature. To correct for ionic strength, many semi-empirical methods are available, none of which is perfect.

How These Models Work

For each equilibrium reaction, an equilibrium constant can be defined. For example, for the reaction

$Cu^{2+} + 4\,Cl^- \rightleftharpoons CuCl_4^{2-}$

The equilibrium constant can be defined as

$$\beta = \frac{[CuCl_4^{2-}]}{[Cu^{2+}]*[Cl^-]^4}$$

Consequently, when the concentrations of free Cu^{2+} and free Cl^- are known, the concentration of $CuCl_4^{2-}$ can be easily calculated as:

$[CuCl_4^{2-}] = \beta * [Cu^{2+}] * [Cl^-]^4$

In fact, the concentrations of free Cu^{2+} and free Cl^- are often NOT known, but what is known are the total concentrations of Cu and Cl in the system. In order to find the speciation, we need to set up a set of mass balance equations needs to be set up, for example:

[total Cu] = [free Cu^{2+}] + [$CuOH^+$] + [$Cu(OH)_2$] + [$Cu(OH)_3^-$] (..) + [$CuCl^+$] + [$CuCl_2$] (..) etc.

[total Cl] = (..)

Each concentration of a complex is a function of the free concentrations of the ions that make it up. So we can say that if we know the concentrations of all the free ions, we can calculate the concentrations of all the complexes, and then we can calculate the total concentrations. A solution to the problem cannot be found by rearranging the mass balance equations, because they are non-linear. What a speciation model does is to repeatedly estimate the free ion concentrations, on each loop adjusting them so that the calculated total concentrations more closely match the known totals. When the calculated and known total concentrations all agree to within a defined precision, the speciation has been calculated. The critical part of the calculation is adjusting the free ion concentrations in a sensible and efficient way to find the solution as quickly as possible. Several more or less sophisticated methods are available to solve this, but usually a Newton-Raphson method is applied.

Influence of Temperature and Ionic Strength

In fact the explanation above is too simple. Equilbrium constants are valid under specific conditions for temperature and ionic strength (for example the standard conditions of 25oC and [endif]--> and need to be converted to the temperature and ionic strength of the system for which speciated is being calculated. It is possible to adapt the equilibrium constants for non-standard temperatures, but this requires knowledge of heat capacity (ΔH) data of each equilibrium. That knowledge is often not available. Constants can be converted from 25°C to other temperatures using the Van 't Hoff-equation:

$$ln(\frac{K_2}{K_1}) = -\frac{\Delta H}{R}(\frac{1}{T_2} - \frac{1}{T_1})$$

where K_1 and K_2 are the constants, T_1 and T_2 the temperatures, ΔH is the enthalpy of a reaction and R is the gas constant.

Equilbrium constants are also valid for one specific value of ionic strength. For conversion from one value of ionic strength to another, many different approaches may be used. This conversion is quite important, because already at relatively low ionic strengths, deviations from ideality become significant, and the activity of a species starts to deviate from its concentration. Hence, the intrinsic, or thermodynamic, equilibrium constants (i.e. constants at a hypothetical ionic strength of zero) are no longer valid and the activity a of ions at non-zero ionic strength needs to be calculated from the concentration and the activity coefficient:

$$a = \gamma * c$$

where γ is the activity coefficient (dimensionless; sometimes also called f) and c is the concentration; a and c are in mol/liter.

The first solution to calculate activity coefficients for non-zero ionic strength was proposed by Debye and Hückel in 1923. The Debye-Hückel theory assumes ions are point charges so it does not take into account the volume that these ions occupy nor the volume of the shell of ligands and/or water molecules around them. The Debye-Hückel gives good approximations, up to circa 0.01 M for a 1:1-electrolyte, but only up to circa 0.001 M for a 2:2-electrolyte. When the ionic strength exceeds these values, the activity coefficients that the Debye-Hückel approximation predicts deviate significantly from experimental values. Many environmental applications require conversions for higher ionic strengths making the Debye-Hückel-equation insufficient. To overcome this problem, many researchers have suggested other methods, like the extended Debye-Hückel-equation, the Güntelberg-equation and the Davies-equation, but also the Bromley-equation, the Pitzer-equation and the Specific Ion Interaction Theory (SIT).

Many programs use the Davies-equation, which calculates activity coefficients γ as follows:

$$log\,\gamma = -0.5079\,z^2\,(\frac{\sqrt{I}}{(1+\sqrt{I})} - 0.3I)$$

where z is the charge of the species and I the ionic strength. Sometimes 0.2 instead of 0.3 is used. Basically all these approaches take the Debye-Hückel-equation as a starting point, and add one or more terms to correct for deviations at higher ionic strengths. Although many of these methods are able to predict the activity of ions fairly well, they are in fact mainly *empirical* extensions without a solid theoretical basis.

Solubility

Most salts have a limited solubility; in several cases the solubility is also important under conditions that occur in the environment. For instance, for $CaCO_3$ the solubility product is 10-8.48, which means that when $[Ca^{2+}] * [CO_3^{2-}] > 10\text{-}8.48$, $CaCO_3$ will precipitate, until $[Ca^{2+}] * [CO_3^{2-}] = 10\text{-}8.48$. But it also works the other way around: if solid CaCO3 is present in a solution where $[Ca^{2+}] * [CO_3^{2-}] < 10\text{-}8.48$ (note the '<'-sign), solid $CaCO_3$ will dissolve, until $[Ca^{2+}] * [CO_3^{2-}] = 10^{-8}.48$. Note that the Ca and CO_3 in the formula here refer to free ions. For example, a 10^{-13} M solution of Ag_2S will lead to precipitation of Ag_2S. The free concentrations of Ag and S are $6.5*10^{-15}$ M and $1.8*10^{-22}$ M resp. (which corresponds with the solubility product of 10-50.12, but the dissolved concentrations of Ag and S are 7.1*10-15 M and 3.6*10-15 M resp., so for S seven order of magnitude higher. This is caused by the formation of S-complexes with protons (HS- and H2S (aq)) and to a lesser extent with Ag.

Complexation by Organic Matter

Complexation with Dissolved Organic Carbon (DOC) is different from inorganic complexation or complexation with well-defined compounds such as acetate or NTA. The reasons for that difference are as follows.

- DOC is very heterogeneous; DOC isolated at two sites may be very different (not to mention the difficulty of selecting isolation procedures).
- Complexation with DOC generally shows a continous range of equilibrium constants, due to chemical and steric differences in neighbouring groups.

- Increased cation binding and/or the ionic strength of the solution change electrostatic interactions among the functional groups in DOC-molecules, which influences the equilibrium constants.
- In addition, changing electrostatic interactions may cause conformational changes of the molecules.

Among the most popular models to assess organic complexation are Model V (1992), VI (1998) and VII (2011), also known as WHAM, written by Tipping and co-authors (Tipping & Hurley, 1992; Tipping, 1994, 1998; Tipping, Lofts & Sonke, 2011). All these models assume that two types of binding occur: *specific binding* and accumulation in the *diffuse double layer*. Specific binding is the formation of a chemical bond between an ion and a functional group (or groups) on the organic molecule. Diffuse double layer accumulation is the accumulation of ions of opposite electrical charge adjacent to the molecule, without formation of a chemical bond (the electrical charge is usually negative, so the ions that accumulate are cations).

For *specific binding*, all these models distinguish fulvic acids (FA) and humic acids (HA) which are treated separately. These two classes of DOC are typically the most abundant components of natural organic matter in the environment – in surface freshwaters, the fulvic acids are typically the most abundant. For each class, eight different discrete binding sites are used in the model. The sites have a range of acid-base properties. Metals bind to these sites, either to one site alone (monodentate), to two sites (bidentate) or, starting with Model VI, to three (tridentate). A fraction of the sites is allowed to form bidentate complexes. Starting with Model VI, for each bidentate and tridentate group three sub-groups are assumed to be present – this further increases the range of metal binding strengths.

Binding constants depend on ionic strength and electrostatic interactions. Conditional constants are calculated in the same way in Model V, VI and VII, as follows:

$$K(z) = K * e^{2wZ}$$

where:

- Z is the charge of the organic acid (in moles per gram organic matter);
- w is calculated by :

$$w = P * log_{10}(I)$$

where:

- *P* is a constant term (different for FA and HA, and different for each model);
- *I* is the ionic strength.

Therefore, the conditional constant depends on the charge on the organic acids as well as on the ionic strength. For the binding of metals, the calculation of the conditional constant occurs in a similar way.

The *diffuse double* layer is usually negatively charged, so it is usually populated by cations, in order to maintain electric neutrality. Calculations for the diffuse double layer are the same for Model V, Model VI and Model VII. The volume of the diffuse double layer is calculated separately for each type of acid, as follows:

$$V_D = \frac{10N_{Av}}{M} * \frac{4\pi}{3} * [(r + \frac{1}{\kappa})^3 - r^3]$$

where:

- N_{Av} is Avogadro's number;
- M is the molecular weight of the acid;
- r is the radius of the molecule (0.8 nm for fulvic acids, 1.72 for humic acids);
- k is the Debye-Hückel parameter, which is dependent on ionic strength.

Simply applying this formula in situations of low ionic strength and high content of organic acid would lead to artifacts (where volume of diffuse layer can be calculated to be more than 1 liter/liter). Therefore, some "tricks" are implemented to limit the volume of the diffuse double layer to 25% of the total.

In case the acid has a negative charge (as it has in most cases), positive and neutral species are allowed to enter the diffuse double layer, just enough to make the diffuse double layer electrically neutral. When the acid has a positive charge, negative and neutral species are present.

The concentration of species in the diffuse double layer is calculated by assuming that the concentration of that species in the diffuse double layer depends on the concentration in the bulk solution and the charge.

In formula:

$$\frac{[X^Z]_{DDL}}{[X^Z]_{solution}} = R^{|Z|}$$

where R is calculated iteratively, to ensure the diffuse double layer is electrically neutral.

Applications

Speciation models can be used for many purposes. Basically, two groups of applications can be distinguished. The first group consists of applications meant to understand the chemical behaviour of any system. The second group focuses on bioavailability.

Chemical Behaviour; Laboratory Situations

Speciation models can be helpful in understanding chemical behaviour in either laboratory situations or field situations. For instance, if you want to add EDTA to a solution to prevent metals from precipitation, the choice of the EDTA-substance also determines the pH of the final solution. Figure 1 shows the pH of a 1 mM solution of EDTA for five different EDTA-salts. This shows that if you want to end up with a near neutral solution, the best choice is to add EDTA as the Na3HEDTA-salt. Adding a different salt requires adding either acid or base, or more buffer capacity, which in turn will influence the chemical behaviour of the solution.

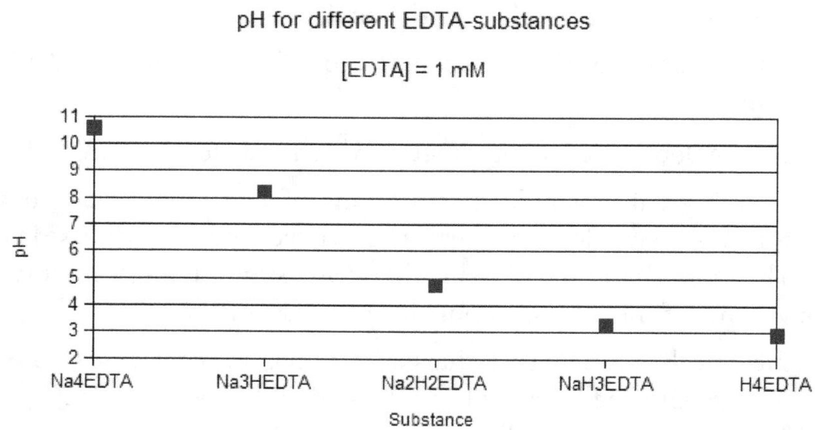

Figure 1. pH of a 1 mM EDTA-solution for different EDTA-salts. Data obtained using speciation program CHEAQS Next.

If you have field measurements of redox potential, speciation models can help to predict whether iron will be present as Fe(II) or Fe(III), which is important because Fe(II) behaves quite different chemically than Fe(III) and also has a quite different bioavailability. The same holds for other elements that undergo redox equilibria like N, S, Cu or Mn.

Phase reactions can be predicted with speciation models, for example the dissolution of carbonate due to the gas solution reaction of CO2. Another example is the speciation in Dutch Standard Water (DSW), a frequently used test medium for ecotoxicological experiments, which is oversaturated with respect to $CaCO_3$ and therefore displays a part of Ca as a precipitate. The fraction that precipitates is very small (less than 2% of the Ca) so it seems unimportant at first glance, but the precipitate induces a pH-shift of 0.22, a factor of almost two in the concentration of free H^+.

Many metals are amphoteric and have therefore a minimum solubility at a moderate pH, while dissolving more at both higher and lower pH-values. This can easily be seen in the case of Al: Figure 2 shows the concentration of dissolved Al as a function of pH (note log-scale for Y-axis). Around pH of 6.2, the solubility is at its minimum. At higher and lower pH-values, the solubility is (much) higher.

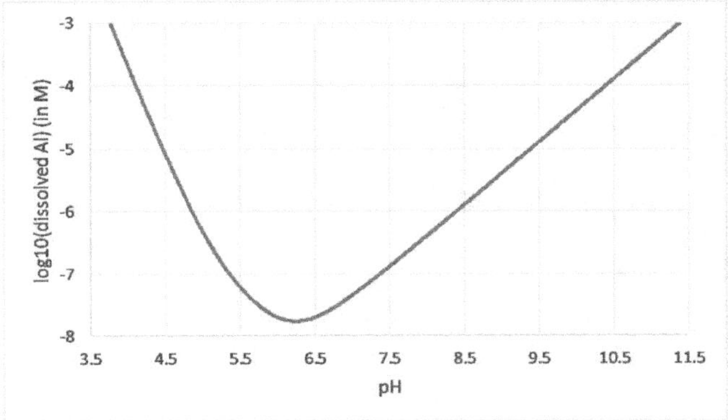

Figure 2. Soluble Al as function of pH. Data obtained using speciation program CHEAQS Next.

Speciation models can also help to understand differences in the growth of organisms or adverse effects on organisms, in different chemical solutions. For example, Figure 3 shows that changes in speciation of boron can be expected only between roughly pH 8 and 10.5, so when you observe a biological difference between pH 7 and 8, it is not likely that boron is the cause. Copper on the other hand (see Figure 4) does display differences in speciation between pH 7 and 8 so is a more likely cause of different biological behaviour.

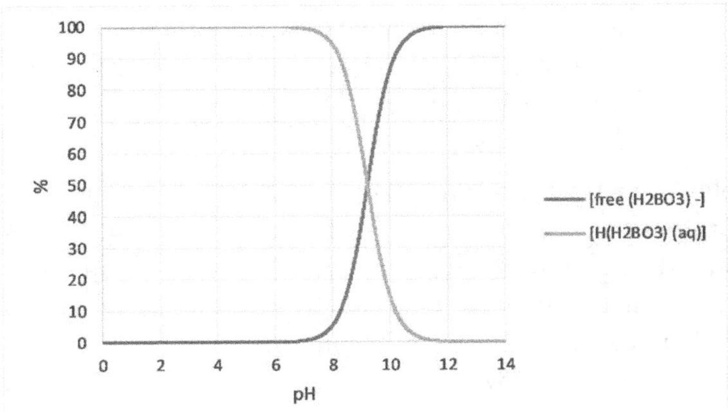

Figure 3. Speciation of B as function of pH; concentration of boron was 1x10-6 M. At higher concentrations, complexes with 2, 3, 4 or 5 B-ions can be formed at significant concentrations. Data obtained using speciation program CHEAQS Next.

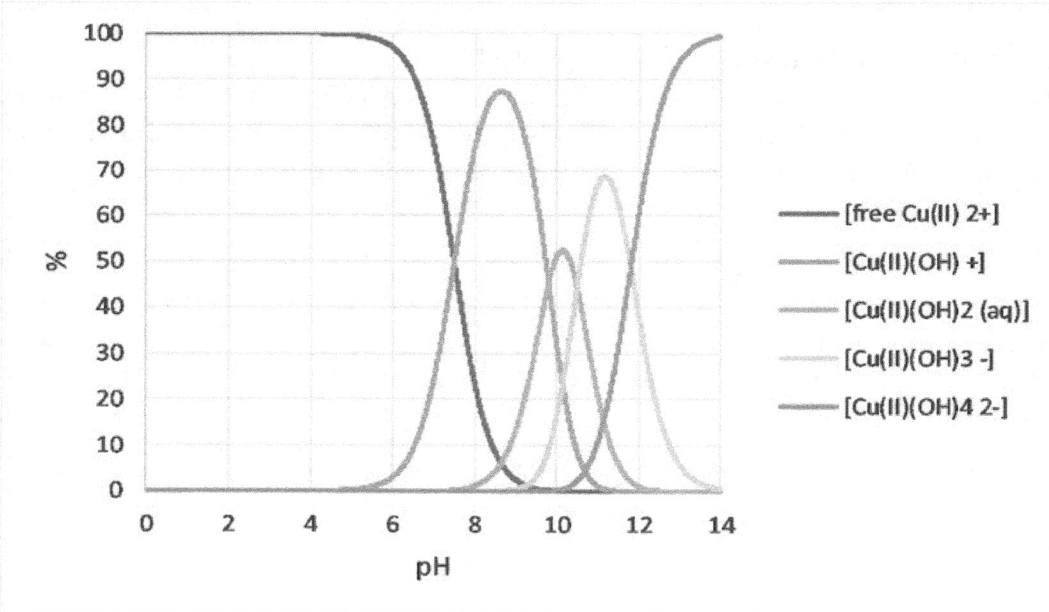

Figure 4. Speciation of copper(II) as a function of pH; concentration of copper was 3x10-8 M. At higher concentrations, complexes with 2 or 3 Cu(II)-ions can be formed at significant concentrations. Data obtained using speciation program CHEAQS Next.

Chemical Behaviour: Field Situations

In field situations, the chemistry is usually much more complex than under laboratory conditions. Decomposition of organisms (including plants) results in a huge variety of organic compounds like fulvic acids, humic acids, proteins, amino acids, carbohydrates, etc. Many of these compounds interact strongly with cations, some also with anions or uncharged molecules. In addition, metals easily adsorb to clay and sand particles that are found everywhere in nature. To make it more complex, suspended matter can contain a high content of organic material which is also capable of binding cations.

For complexation by fulvic and humic acids, Tipping and co-workers have developed a unifying model (Tipping & Hurley, 1992; Tipping, 1994, 1998; Tipping, Lofts & Sonke, 2011). The most recent version, WHAM 7 (Tipping, Lofts & Sonke, 2011), is able to predict cation complexation by fulvic acids and humic acids over a wide range of chemical circumstances, despite the large difference in composition of these acids. This model is now incorporated in several speciation programs.

Suspended matter may be of organic or of inorganic character. *Inorganic matter* usually consists of (hydr)oxides of metals, such as Mn, Fe, Al, Si or Ti, and clay minerals. In practice, the (hydr)oxides and clays occur together, but the mutual proportions may differ dramatically depending on the source. Since the chemical properties of these metal (hydr)oxides clays are quite different, there is a huge variation in the chemical properties of inorganic suspended matter in different places and different times. As a consequence, modeling interactions between dissolved constituents and suspended inorganic matter is challenging. Only by measuring some properties of suspended inorganic matter, can modeling be applied successfully. For suspended organic matter, the variation in properties is also large and modelling is challenging.

Bioavailability

Speciation models are useful in understanding and assessing the bioavailability of metals and other elements in test media. Test media often contain substances like EDTA to keep metals in solution. EDTA-complexes in general are not bioavailable, so in addition to keeping metals in solution they also change their bioavailability. Models can calculate the speciation and help you to assess what is actually happening in a test medium.

An often forgotten aspect is the influence of CO_2. CO_2 from the ambient atmosphere can enter a solution or carbonate in solution (if in excess over the equilibrium concentration) can escape to the atmosphere. The degree to which this exchange takes place, influences the pH of the solution as well as the amount of carbonate that stays in solution (carbonates are often poorly soluble).

Similarly, in field situations models can help to understand the bioavailability of elements. As stated above, the influence of DOC can nowadays be assessed properly in many situations, the influence of suspended matter remains more difficult to assess. Nevertheless models can deliver insights in seconds that otherwise can be obtained only with great difficulty.

Models

There are many speciation programs available and several of them are freely available. Usually they take a set of total concentrations as input, plus information about parameters such as pH, redox, concentration of organic carbon etc. Then the programs calculate the speciation and present them to the user. The equations cannot be solved analytically, so an iterative procedure is required.

Although different numerical approaches are used, most programs construct a set of non-linear mass balance equations and solve them by simple or advanced mathematics. A complication in this procedure is that the equilibrium constants depend on the ionic strength of the solution, and that this ionic strength can only be calculated when the speciation is known. The same holds for the precipitation of solids. The procedure is shown in Figure 5.

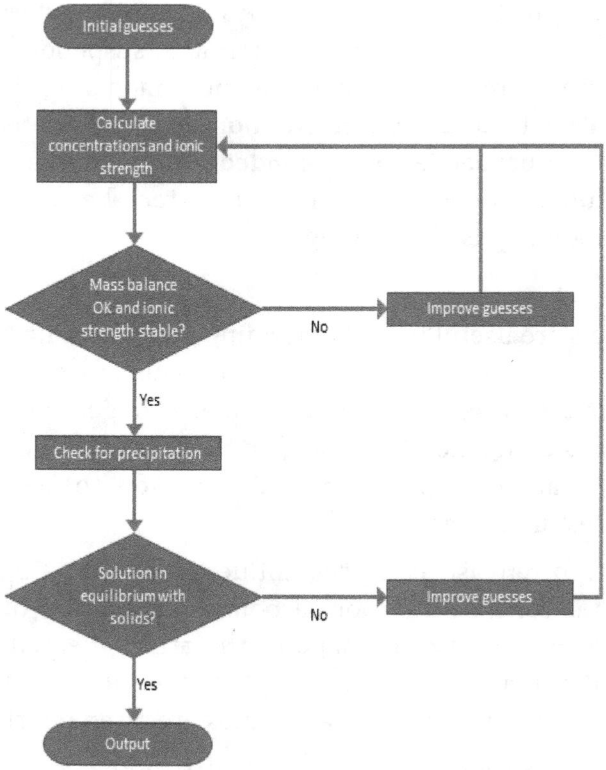

Figure 5. Typical flow diagram of a speciation program.

Limitations

For modeling speciation, thermodynamic data is needed for all relevant equilibrium reactions. For many equilibria, this information is available, but not for all. This hampers the usefulness of speciation modeling. In addition, there can be large variations in the thermodynamic values found in the literature, resulting in uncertainty about the correct value. A factor of 10 between the highest and lowest values found is not an exception. This of course influences the reliability of speciation calculations. For many equilibria, the thermodynamic data is only available for the standard temperature of 25°C and no information is available to assess the data at other temperatures, although the effect of temperature can be quite strong. Also ionic strength has a high impact on equilibrium 'constants'; there are many methods available to correct for the effect of ionic strength, but most of them are at best semi-empirical. Simonin (2017) recently proposed a method with a solid theoretical basis; however, the data required for his method are available only for a few complexes so far.

More fundamentally, you should realize that speciation programs typically calculate the equilibrium situation, while some reactions are very slow and, more inportant, nature is in fact a very dynamic system and therefore never in equilibrium. If a system is close to equilibrium, speciation programs can often make a good assessment of the actual situation, but the more dynamic a system, the more care you should take in believing the programs' results. Nevertheless it is good to realise that a chemical system will always move towards the equilibrium situation, while organisms may move them away from equilibrium. Phototrophs are able to move a system away from its equilibrium situation whereas decomposers and heterotrophs generally help to move a system towards its equilibrium state.

References

Simonin , J.-P. (2017). Thermodynamic consistency in the modeling of speciation in self-complexing electrolytes. Ind. Eng. Chem. Res. 56, 9721-9733.

Tipping, E., Hurley, M.A. (1992). A unifying model of cation binding by humic substances. Geochimica et Cosmochimica Acta 56, 3627 - 3641.

Tipping, E. (1994). WHAM - A chemical equilibrium model and computer code for waters, sediments, and soils incorporating a discrete site/electrostatic model of ion-binding by humic substances. Computers & Geosciences 20, 973 - 1023.

Tipping, E. (1998). Humic Ion-Binding Model VI: An Improved Description of the Interactions of Protons and Metal Ions with Humic Substances. Aquatic Geochemistry 4, 3 - 48.

Tipping, E., Lofts, S., Sonke, J.E. (2011). Humic Ion-Binding Model VII: a revised parameterisation of cation-binding by humic substances. Environmental Chemistry 8, 228 - 235.

Further Reading

Stumm, W., Morgan, J.J. (1981). Aquatic chemistry. John Wiley & Sons, New York.

More, F.M.M., Hering, J.G. (1993). Principles and Applications of Aquatic Chemistry. John Wiley & Sons, New York.

3.8.3. Modeling Exposure at Ecological Scales

In Preparation

www.ingramcontent.com/pod-product-compliance
Lightning Source LLC
LaVergne TN
LVHW080056220926
848484LV00007BA/1741
* 9 7 8 1 7 8 7 1 5 1 2 8 4 *